India is one of the most populous countries with nea~~ ~~ global population out of which a good 70% resides across 730,000 villages. Thus, the true growth of India in any measure depends on the growth and development of rural villages and the empowerment of the villagers, leading to sustained and resilient lifestyles. The book *Infrastructure for Smart Villages* has been brought into reality at an interesting juncture of the rapid transformation of rural villages where new approaches and underlying considerations are being highlighted for supporting sustainable development practices in rural settings. Contrasting to the traditional development models, this book offers a promising guide for professionals, policymakers, and other key stakeholders on planning, development, and operations including materialisation of the UN's SDGs in context-specific implementation

Prof V K Vijay, *IREDA Chair Professor, Centre for Rural Development and Technology, National Coordinator – Unnat Bharat Abhiyan program of Ministry of Education, Govt of India, Coordinator – Biogas Development and Training Centre of MNRE, Govt of India Indian Institute of Technology Delhi, New Delhi, India*

Infrastructure for Smart Villages

This book intends to initiate a fresh articulation of need-based infrastructure provisions in rural contexts. Departing from the conventional theories and practices of infrastructure planning and development applied in urban settings, the book presents a comprehensive suite of technical and non-technical indicators that rationalise fit-for-purpose planning, development, and operations of rural infrastructure. Drawing from global practices in public and private sectors and research-based evidence, a distinctive argument is put forward for promoting location-specific infrastructure development from effectiveness, practicality, affordability, and sustainability perspectives. The argument encompasses wider social, cultural, and economic contexts that are unique to rural settings and the book highlights a clear roadmap of how the UN's sustainable development goals (SDGs) are at the core of developing rural communities with necessary infrastructure provisions that are purpose-built, affordable, risk-averse, and resilient.

This book will provide an overview of some of the little-understood and sometimes counter-intuitive best practices on rural infrastructure and value-based priorities that have emerged in uplifting rural communities in developing economies over the last 30 years. Drawing from the global literature and practice-based evidence across a complete spectrum of relevant disciplines, this book will provide readers with a clear articulation of the innovative ideas around harnessing rural potential, and empowering rural communities with added support in growth and progressive development in the context of interconnected infrastructure systems and improved living standards. It is key reading for development, planning, and infrastructure courses as well as professionals and researchers involved in international development, aid, and provision in rural areas.

Dr Hemanta Doloi is an Associate Professor in the Construction Management discipline and the Project Director of the Smart Villages project at the Faculty of Architecture, Building and Planning of the University of Melbourne.

Infrastructure for Smart Villages

Hemanta Doloi

Routledge
Taylor & Francis Group

LONDON AND NEW YORK

First published 2025
by Routledge
4 Park Square, Milton Park, Abingdon, Oxon OX14 4RN

and by Routledge
605 Third Avenue, New York, NY 10158

Routledge is an imprint of the Taylor & Francis Group, an informa business

British Library Cataloguing-in-Publication Data
A catalogue record for this book is available from the British Library

ISBN: 978-1-032-62229-3 (hbk)
ISBN: 978-1-032-61857-9 (pbk)
ISBN: 978-1-032-62232-3 (ebk)

DOI: 10.1201/9781032622323

Typeset in Times New Roman
by MPS Limited, Dehradun

Contents

3 Infrastructure and economy in rural and regional context　　　**44**

Acknowledgements

The author wishes to express profound gratitude and sincere appreciation to the Faculty of Architecture, Building and Planning at the University of Melbourne for facilitating the Smart Villages research across transnational institutions and developing shared capacities in sustainable rural construction, an area of critical need. A sincere appreciation goes to the Government of Assam, India, for providing generous funding support to initially establish the research collaboration between the University of Melbourne, Assam Engineering College, and Indian Institute of Technology (IIT) Guwahati and enable the Smart Villages research under the leadership of the author. The author extends his sincere thanks to Late Tarun Gogoi, the past Honourable Chief Minister of Assam and Dr Himanta Biswa Sarma, the current Honourable Chief Minister of Assam, for their unconditional support and encouragement to continue this important area of research and for creating an impact over 86% rural population in the State of Assam. The author is also thankful to a large number of people from the Government of Assam who were closely involved with their unwavering support to conduct the empirical research taking Assam as the test bed over four years 2016–2020. The vision and mission of these visionary leaders for making a real difference in rural communities in the state through scientific research and development with a world-class institution is highly significant.

A special thanks goes to Prof Julie Willis, Dean of the Faculty of Architecture Building and Planning for her encouragement and support in Smart Villages research in the Faculty. The leadership and facilitation of a few colleagues from the Faculty and in particular from the Construction Management discipline are highly valuable.

The author is highly appreciative of two distinguished colleagues, Professor Mohan Kumaraswamy, Editor-In-Chief of the Emerald Journal of *Built Environment Project and Asset Management (BEPAM)*, and Professor George Ofori, London South Bank University, UK for their continuous support and encouragement in Smart Villages research. Continuous mentorship and encouragement also received from a few distinguished colleagues from the IIT Delhi, IIT Madras, IIT Tirupati, IIT Guwahati, and Assam Engineering College, Assam, India are truly inexpressible by words.

A very special thanks goes to Dr. Mohammad Karamloo for his support in gathering the relevant literature in the early part of this book project.

The author expresses his sincere gratitude to the Non-Resident Indian (NRI) Welfare Society of India for awarding him two consecutive international awards, the "Nelson Mandela Leadership Award 2023" and "Hind Rattan (Jewel of India) Award 2024" recognising his outstanding services, contributions, and achievements in research and teaching and improving the human lives globally.

The author is particularly thankful to his wife Mrs Monjita Dutta Doloi and the two most beautiful young boys, Rishov and Rahul for their constant encouragement and push to complete this book without taking time off even during the holiday periods. Last but not least, the publisher deserves special thanks for their kind support and interest in the project.

About the author

Dr Hemanta Doloi is an Associate Professor in the Construction Management discipline and the Project Director of the Smart Villages project at the Faculty of Architecture, Building and Planning of the University of Melbourne. He is the director of the Smart Villages Lab (SVL) which focuses on data-driven research in the area of critical needs "Sustainable Rural Construction and Development" under the auspice of the Smart Villages program. He is also the founding Chair of the International Conference on Smart Villages and Rural Development (COSVARD), founding Convener of the Global Seminar Series on Smart Villages (GSSV), and the Mission Advocate of the Sustainable Planet Initiative (SPI).

He has conducted significant conceptual and empirical research and advanced the Construction Management discipline in Australia and internationally focusing on four key areas: (1) Project management, (2) infrastructure planning and policy, (3) construction economics and management, and (4) rural construction and smart villages. He leads the trans-disciplinary research in Smart Villages for developing sustainable solutions for affordable housing and infrastructure systems, generating new theories for education and governance, and empowering rural communities. Focusing on "THE 40 PERCENT" global population who still resides in remote villages under extremely rural and under-developed conditions in the world, his Smart Villages research is particularly contributing to local empowerment and thereby stemming rural migrations into overly crowded cities. His evidence-based Smart Villages research demonstrates "Urban-Rural Share" as being the most sustainable way of developing "THE 40 PERCENT" rural communities instead of the current race against time to close the "Urban-Rural" gap.

In recognition of contributions, services, and achievements in his academic career particularly making positive impacts on human lives through his research, Hemanta was awarded the "Nelson Mandela Leadership Award 2023" and "Hind Rattan (Jewel of India) Award 2024" by the Non-Resident Indian (NRI) Welfare Society of India. Hemanta also won the Australian Institute of Quantity Surveyor's "Infinite Value Award in Teaching and

Research" – recognising the excellence of scholarships impacting the wider community in the profession. Hemanta authored or co-authored three books in the past and has published widely in leading peer-reviewed journals and conferences internationally. He serves as an Associate Editor, Guest Editor, and editorial board member in several international journals of high repute. Hemanta is widely consulted in government and Industry sectors globally.

Foreword

This is a timely cutting-edge book that may have also been called an "out of the box" book a few years ago. While well-positioned at a deeper cutting-edge now, Hemanta Doloi has now brought the theme of this book "into the picture" or "inside the box" of many in our built infrastructure community, with publications, conferences, and journal special issues emanating from the research and development he leads at the Smart Villages Lab at the University of Melbourne. I first had the pleasure of interacting with Hemanta at IIT Madras when we were both visiting there around 2016 and have seen him as a rising star in our field, also aware of his laudable efforts in setting up and developing the Smart Villages Lab. Indeed, I have been fortunate to play some minor roles in the first journal special issue that he edited on this theme and his two most recent conferences on this smart villages and rural infrastructure too. Therefore, I have had many opportunities to appreciate and learn from the vast strides taken in this critical but hitherto relatively neglected field.

The book is timely at a time when the World is struggling to meet the UN's sustainable development goals (SDGs). Indeed, these may seem to have become moving targets, given goal posts shifted by relentlessly progressing climate change, punctuated by both natural and human-induced disasters. Hemanta's valuable, research, findings, and recommendations on ways forward, as neatly packaged in this new book, broaden the traditional focus area in pursuing SDGs: from the urban to rural. This includes bringing smart villages into the picture alongside smart cities and the infrastructure networks linking them too.

Infrastructure for Smart Villages presents readers with a highly significant and original body of knowledge on one of the woefully under-researched but increasingly important topics of creating purpose-based and multi-functional rural infrastructure. The construction industry is a key contributor to the rising challenges of climate change and global warming arising from careless human interventions and thoughtless disruptions of nature. Recent trends of rapid urbanisation especially in emerging economies are certainly not helping to contain the emissions of greenhouse gases while "developed" economies are also reluctant to control, leave along "let go" of the main polluters.

While a good 40% world's population still lives in rural settings, urbanisation is attracting rural youth at unprecedented rates, rather than engaging them in conceiving and operationalising sustainable development solutions in rural areas. Therefore, empowering prime human capital in rural areas to engage in such sustainable development would also stem the flow of misdirected urban migration. The "right" (appropriate) infrastructure is a precursor to sustainable development anywhere, yet the pace of infrastructure development in rural settings is not as high as it should be, which could also be due to a lack of attractive investment opportunities for potential private investors.

Essential knowledge and investment frameworks including relevant governance models for purpose-built rural infrastructure systems are needed and provided between the covers of this book. The book also helps to rethink and recalibrate approaches toward rural infrastructure needs and to empower rural communities to aspire a distinctive rural lifestyle based on their own context that could be more rewarding and feel better in some ways than that of urban communities. This timely book puts together a succinct argument on how evidence-based approaches can support planning and development of context-specific and purpose-built infrastructure systems in rural settings. Drawing from his research and findings in infrastructure planning and procurement in general and Smart Villages in particular, Hemanta has developed and now made available, a vast body of new knowledge articulating the bottom-up and participatory governance model in developing Smart Villages. This model also leads to a fresh Urban-Rural Share approach, if not paradigm towards striking the essential, but hitherto lacking, broad-based balance in targeted global sustainability outcomes. The book also showcases alternative development models as a driving force for generating transformative impact of rural infrastructure systems which are at the core of context-specific implementation of the UN's SDGs in uplifting the rural communities towards a self-sustained resilient and distinctive rural lifestyle. This book should serve as a major reference point and enabler for practitioners, academia, rural leaders, and policymakers who are interested and/or involved in rural development worldwide.

Mohan M. Kumaraswamy
Honorary Professor, The University of Hong Kong, Hong Kong SAR,
Honorary Professor, University of Moratuwa, Sri Lanka, Founding Director,
Centre for Innovation in Construction & Infrastructure Development,
Joint Co-ordinator, CIB Working Commission on Public-Private Partnership,
Editor-in-Chief, Journal of Built Environment Project and Asset Management
(BEPAM), Editor-in-Chief, World Scientific Book Series on Domain-Specific
Bodies of Knowledge in Project Management

Foreword

There is much coverage of the proportion of the population living in the urban areas. Attention is drawn to the growth of this figure, and focus is put on the need to address the needs of the new additions to the urban population. There is a large volume of work on the need for improving the management of the urban areas in order to address problems such as shortage of housing and poor quality of existing houses at the same time as spreading urban sprawl; inadequate infrastructure and services causing traffic congestion together with air pollution and problems of public health such as poor sanitation. The belief that cities are the engine of economic, which is backed up with data showing that close to three-quarters of national income and employment has been used to justify the concentration of national plans, policies on cities. This focus is evident in names such as the "New Urban Agenda", the global programme for sustainable physical development (although that programme does give some coverage to rural development).

The situation of the population in the rural areas is not adequately highlighted, and even less satisfactorily addressed. One would have thought that the "left behind population" would be front and centre of the action in sustainable development, at least to reduce the outflow of humans. The conditions in the rural areas are also increasingly worsening as a result of climate change and the increasing frequency of natural disasters. Action here would likely be simpler and cheaper to accomplish than that in the cities; it is likely to go a longer way in terms of coverage by proportion of the population. It should also be pointed out that these are also citizens of the country, entitled to an improvement in their quality of life. The population should also be provided with the most modern facilities. Thus, Smart Villages is a subject for research. It is neither a misnomer nor the opposite of Smart Cities. Each of them has its features and desirable characteristics.

Infrastructure for Smart Villages is a relevant, timely, and welcome subject for research and articulation. This book sets out to consider the determination, planning, design, and delivery of infrastructure projects in the rural context. It argues that the concepts, principles, and practices of infrastructure planning, development, and operation that are applied in urban settings are ill-suited to rural areas as their contexts are different. The book considers the dynamics of infrastructure provisions across various sectors such as health, transport, water, sanitation, waste, and recreational facilities. It presents an alternative approach based on the author's research on the smart village concept which will deliver effectiveness, practicality, affordability, sustainability, and resilience on the projects and in their products. It considers the assessment of the needs and requirements and definition of the priorities that arise from the physical, demographic, economic, social, and cultural changes occurring in rural areas.

It is time to focus on "THE 40 PERCENT", which has been referred to by other names, some of which are not always complimentary, such as "The Bottom Billions" or "The Urban Poor", and addressing their infrastructure needs in ways which take cognisance of the contexts in which they live and work and involves them in the entire process is a good first step. It is pertinent that Hemanta Doloi has drawn on many years of research on Smart Villages to produce this book. This will be a useful guide for the planning and management of programmes for the sustainable development of rural areas in countries at all levels of economic progress. It is reasonable to surmise that the ideas here will also contribute to the mainstream body of knowledge of development and project management.

It is time for "THE 40 PERCENT" to characterise a movement which seeks to improve the quality of life of the world's population in the rural areas of all countries. It is my hope that Smart Cities and Smart Villages will be studied together to find synergistic solutions for sustainable development for the benefit of all in society.

George Ofori
Professor
Sustainability Lead for LSBU Group,
London South Bank University (LSBU), London, UK

Preface

Having spent a long research career in construction economics, project procurement, infrastructure planning and delivery, and recently rural construction and creation of Smart Villages targeting over "THE 40 PERCENT" rural population in the globe, the author is extremely delighted to bring this book into a reality. In the advent of climate change and global warming, vastly due to man-made destruction to nature, the Smart Villages' research towards alternative and sustainable development models in modernising this significant section of rural communities on the planet presents an unprecedented opportunity and contributes towards the reduction of greenhouse gas emissions. Infrastructure is the backbone of the development of rural communities. Given the emergence of the topic and high carbon footprints in infrastructure projects, alternative sustainable models of infrastructure planning and development have become an important consideration globally. Addressing these challenges, for the first time, this book comprehensively puts together the key considerations required for creating purpose-built infrastructure provisions keeping the communities at the core of decision-making in projects. The book should be of interest to practitioners and academics in policy institutes, intergovernmental organisations, and universities.

The author views this book "*Infrastructure for Smart Villages*" as a natural extension of the first conceptual book "*Planning, Housing and Infrastructure for Smart Villages*" published by Routledge in February 2019, but with an in-depth inquiry on the topic of infrastructure covering all aspects of rural development in emerging economies. Referring to the prevailing practices of rural development, the book particularly delves into the ways to combine the vernacular knowledge and emerging factors necessary for creating functional and inclusive infrastructure systems in Smart Villages. The second book "Affordable Housing for Smart Villages" also published by Routledge in January 2020 provided the necessary references and contexts in the natural progression of this infrastructure book.

This book intends to initiate a fresh articulation of need-based infrastructure projects in a rural context. Departing from the conventional theories and practices of infrastructure planning and development applied in urban settings, the book presents a comprehensive suite of technical and non-technical indicators that rationalise fit-for-purpose planning, development, and operations of rural infrastructure. Drawing from the global practices in public and private sectors and research-based evidence, the author puts forward a distinctive argument for promoting location-specific infrastructure development from effectiveness, practicality, affordability, and sustainability perspectives. The argument encompasses wider social, cultural, and economic contexts that are unique to rural settings. It presents a clear roadmap of how the UN's sustainable development goals (SDGs) are at the centre of developing rural communities with necessary infrastructure provisions that are purpose-built, affordable, risk-averse, and resilient.

Highlighting the significance of context-specific rural infrastructure, the author shows the ways of assessing the needs and requirements and redefining the priorities that arise from the sweeping changes occurring in rural areas. Some of the natural changes experienced in rural include continuing out-migration and demographic shifts, uneven distribution of wealth from emerging rural markets, land economics and loss of fertile lands, changing aspirations, and cultural influences in construction. In short, the dynamics of infrastructure provisions across all sectors (such as health, transport, water, sanitation, waste, public and recreational facilities, etc.) can be better assessed with a data-driven upward approach which needs to be carefully assessed with the balance of capital in rural areas and values being created in the localised settings. The book conceptualises the infrastructure along a spectrum of rural lifestyles that is interlaced with cultural and social values integral to rural livelihoods at both personal and community scales.

The book is developed in eight integrated chapters. For those residing in rural areas, having access to functional infrastructure is kind of a luxury. Any work in progress with incremental quality improvement along with careful management of irregular resource flows is an important consideration for progressively developing and meeting the infrastructure needs of the rural communities. Such a practice is often nestled alongside broader socio-economic, cultural, and political contours of rural community life which need to be reflected in rural development policies. Rural infrastructure schemes also need to incorporate the influence of emerging technologies and digital illiteracies across rural communities.

The book should be an essential tool in most institutions that work with a focus on rural development, rural infrastructure, social value, and heritage, education policies, or planning studies. It should serve as a single point of reference for developing the theoretical underpinnings of any policy planning and academic programs, especially within the technical institutes in the developing economies in the context of developing the rural infrastructure and community by leveraging the effective use of scarce resources.

1 Infrastructures, society, and community

1.1 Introduction

Appropriate provisions of infrastructure for the effective functioning of society are a fundamental need for any community. Yet, providing basic infrastructure facilities remains a major challenge across many countries especially developing countries across the globe. With limited funds coupled with a lack of access to skills, expertise, resources, and experiences, developing countries are unable to realise the full potential of their population across sectors such as education, job-creation, productivity growth, self-sufficiencies, reduction of vulneraries, and increasing resiliency in the society. Leading organisations such as the International Labour Organisation (ILO) strive to formulate necessary strategies and tools for supporting the development of basic rural infrastructure provisions such as roads, bridges, railway tracks, waste supply, sanitation, irrigation, education, health, and retail market; yet, successful implementation of such strategies and delivering context-specific solutions are quite limited among the rural communities.

In the advent of rapid urbanisation in developing economies and the race against time for uplifting remote and rural communities with basic infrastructure needs, access to the context-specific body of knowledge is highly crucial. However, due to a lack of rural-centric knowledge, the extension of urban-centric practices for rolling out the rural infrastructure becoming the norm across many countries. As a result, the fundamental principles of localised considerations such as local economy, local materials, and resources, decentralised infrastructure network, social enterprises, etc. are compromised which in turn contributes to inefficiencies in infrastructure operations and maintenance.

The smart villages research which strives to develop alternative and sustainable models targeting over 40% of rural communities in the globe is at the core of this book. As informed by the research, need-based infrastructure provisions in the rural context are highly relevant for enabling rural communities to fast-track the development ladder. Conventional theories and prevailing practices of city-centric infrastructure planning and development do not always lend the best capacity for rural planning and development. While decision-making on urban infrastructure projects is influenced by financial parameters, the decisions on

DOI: 10.1201/9781032622323-1

rural infrastructure are predominately linked to the community-centric requirements aligning to the social, and cultural including myriad non-financial parameters. Informed by the field research, this book presents a comprehensive suite of technical and non-technical indicators that rationalise fit-for-purpose planning, development, and operations of rural infrastructure. The core of the argument for need-based infrastructure development is the in-depth understanding and appreciation of the intrinsic characteristics and interconnectedness of infrastructure systems concerning society and community. Drawing from the evidence and reflecting on the shortcomings of the rapid urbanisation of rural communities across emerging economies, a distinctive argument is put forward for promoting community-led infrastructure development practice from effectiveness, practicality, affordability, and sustainability perspectives. The argument encompasses wider social, cultural, and economic contexts that are unique in rural conditions, ensuring the infrastructure interventions are responsive to the fit-for-purpose of the target communities, and enabling sustained progressions in the farm-based rural economies.

The book highlights a clear roadmap of how the UN's Sustainable Development Goals (SDGs) are at the core of developing rural communities with necessary infrastructure provisions that are purpose-built, affordable, risk-averse, and resilient (United Nations, 2015). The 17 SDGs include G1: No poverty; G2: Zero hunger; G3: Good health and well-being; G4: Quality education; G5: Gender equity; G6: Clean water and sanitation; G7:Affordable and clear energy; G8: Decent work and economic growth; G9: Industry, innovation, and infrastructure; G10: Reduced inequalities; G11: Sustainable cities and communities; G12: Responsible consumption and production; G13: Climate action; G14: Life below water; G15: Life on land; G16: Peace, justice, and strong institutions; and G17: Partnerships for the goals. While developing the contents of the book, an attempt is being made to make a clear reference to the appropriate goals for highlighting the importance and significance of relevant infrastructure and supporting the same. The remainder of the chapters will articulate how the SDGs could be promoted by having appropriate infrastructure provisions in rural settings.

Highlighting the significance of purpose-built rural infrastructure, the book shows the ways of assessing the needs and requirements and redefining the priorities that arise from the sweeping changes occurring in rural areas (Prud'homme 2004). Some of the natural changes experienced in rural communities include rapid urbanisation, loss of fertile land, irregular planning and development, lack of infrastructure provisions, lack of equity and accessibility, degradation of cultural heritage, artifacts, and overall social values. In short, interrelations of infrastructure provisions across all sectors (such as health, transport, water, sanitation, waste, public and recreational facilities, etc.) can be better assessed with a data-driven upward approach which needs to be carefully assessed with the balance of capital in rural areas and values being created in the localised settings. The book conceptualises the infrastructure along a spectrum of rural lifestyles about cultural and heritage sensitivity, social norms, and livelihoods at personal and community levels.

The book contains eight integrated chapters. One of the key intents across all the chapters is to highlight the significance and importance of accessibility and integration issues while rolling out an incremental or phased-based development of the overall infrastructure systems. This is because having access to functional infrastructure is kind of a luxury for many residing in rural settings and thus the infrastructure systems must be able to latch the community at both micro and macro levels. Infrastructure being the backbone of the development of the rural community, the policies must be centred around the broader socio-economic, cultural, and political contours of the community. Leveraging digital and information technology increases the accessibility of the infrastructure systems from both functional and operational aspects.

1.2 Definition of infrastructure

There are different ways to define the term "infrastructure". One way to do so is to describe this name by its characteristics as depicted in Figure 1.1. For example, the prominent economist, Hirschman, used the idea of "social overhead capital" and defined the infrastructure as "capital that provides public services" (Hirschman 1958). In other words, two intrinsic features of the infrastructure in this definition are "publicness" as well as being "durable" enough to provide services (capital) (Fourie 2006). However, the former part should not be mixed up with the notion of being publicly owned as this could easily be debated in cases such as privately-owned telecommunication networks, or energy distributors. In this regard, some economists believe that the notion of publicness should be defined as strong public involvement (Rietveld and Bruinsma 1996). Other intrinsic characteristics of a typical infrastructure type may include non-rivalry, non-excludability, indivisibility, non-substitutability, immobility, polyvalence, affordability, accessibility, and publicness. However, not all types of infrastructures may need or require to inherit all these characteristics (Rietveld and Bruinsma 1996). Therefore, defining the infrastructure through its characteristics could be a cumbersome and debatable task. To remedy this difficulty, another approach has been taken by other researchers in which the specific infrastructure is labelled by its application such as energy-related infrastructures, transport infrastructures, etc. though this approach could be too subjective based on the perception of the researchers who are using the taxonomy (Fourie 2006).

According to the Longman dictionary of contemporary English, infrastructure is defined as "the basic systems and structures that a country or organisation needs to work properly, for example, roads, railways, banks, etc." (Summers 2003). However, this definition is prone to be a misleading explanation as the infrastructure definition should encompass different dimensions. In this regard, infrastructures could be categorised into two main categories, hard and soft infrastructures. Hard infrastructures which typically refer to ports, roads, bridges etc. do not always represent the underlying intents of the infrastructure systems about the basic functions being served or required for the effective functioning of a system. However, such nomenclature of the infrastructure types needs to

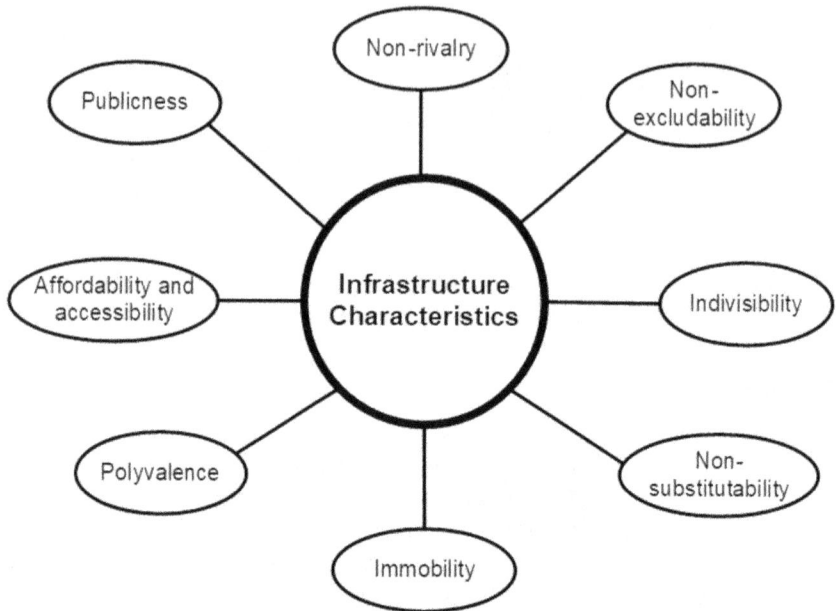

Figure 1.1 Characteristics of a typical infrastructure.

comprehend soft infrastructure functionalities such as skills, education, socio-demographic properties, hope, interpersonal relations, etc. for the system to work effectively. For instance, a governmental body could build a school in a remote area, however, it is impossible to use this hard infrastructure properly unless the government could offer proper incentives to encourage skilled teachers to pursue their careers in such locations.

1.3   Hard versus soft infrastructures and their significance

The definition of hard and soft infrastructures could be tricky as in different contexts, they may be defined with different explanations. For instance, Kavanagh and his co-workers (Kavanagh, Shiell et al. 2020) reviewed the health-related literature and mentioned that the soft infrastructure could encompass hope, trust, safe spaces, cultural symbols, heritage and social values, etc. However, Evans and O'Brien (Evans and O'Brien 2015) defined the soft infrastructure as everything that hard infrastructure is not. They explained that there is no general agreement on the definition of soft infrastructure and added that this concept could include all organisations or institutions that are needed to improve or maintain the socio-cultural or health-related standards of a country. In the context of society, this definition might be referring to public literacy or welfare and health (Evans and O'Brien 2015). Surprisingly, in the domain of computer science, the soft infrastructure denotes networking, the technology of communication, and all the systems with the internet (Evans and O'Brien 2015).

However, one of the good definitions of hard and soft infrastructures is presented by Gibson (2017) who described hard infrastructure as the physical resources and systems that provide different services to the people, community, and organisations while mentioning soft infrastructure as the unseen type of infrastructure which could be either as broad as communities' social support or in the form of institutions that are managing the physical (hard) infrastructures (Chirisa and Nel 2021) (Images 1.1–1.3).

Image 1.1 A significant piece of social infrastructure in a rural village in Assam known as "Chikon Atta Than", a pilgrimage place for regular social congregations.

Image 1.2 Gate to the pilgrimage place in Assam with sculptures of divine images "Mahapurus Srimanta Sankardev", generating added respect among the devotees.

Image 1.3 An Architectural view highlighting the scale of the "Chikon Atta Than", a highly auspicious place for devotes spanning across the state of Assam.

1.4   Physical/economic versus social infrastructures and their significance

Figure 1.2 shows a broad classification of infrastructure types in a generalised context. As seen, infrastructure is usually divided into two broad areas, economic infrastructure and social infrastructure. Economic infrastructure is usually linked to economic development and productivity of a country which then contributes to the Gross Domestic Product (GDP) as a measure of the economic health of a nation. Social infrastructure is generally linked to the

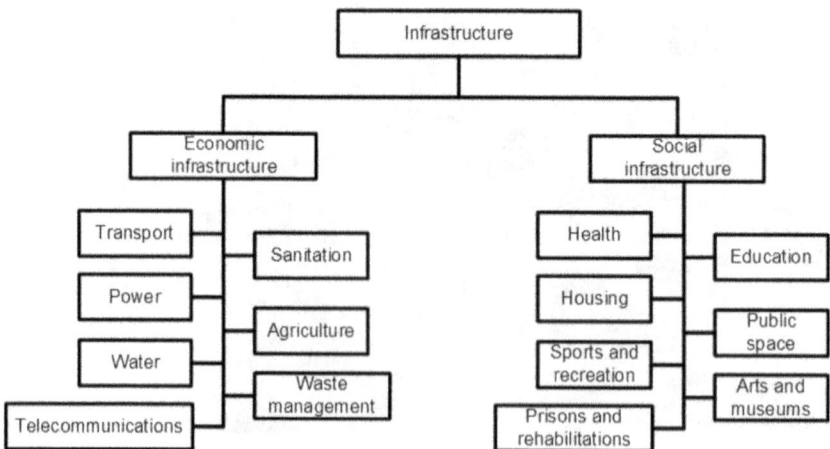

Figure 1.2 Infrastructure types and classifications.

social well-being of the community which contributes to the overall societal support and growth.

As found in the mainstream literature, physical or economic infrastructure refers to the structures that can shape a place (Brown 2021). This could include either built objects such as roads, bridges, and buildings or natural objects (Brown 2021). However, social infrastructures could be defined as a place for interactions to take place (Brown 2021). There is an ambiguity within the literature mixing up the definitions of social capital and social infrastructure. However, social capital could be defined either as trust between the society members (Brown 2021), or as providing emotional and instrumental support within a society to recover from a disaster at a household or community level (Aldrich 2010, Aldrich and Meyer 2014). Another concept is the definition of the so-called entrepreneurial social infrastructure which according to the definition of Flora and Flora (1993) should have three main elements including diversity to enable constructive criticism and disagreement, mobilisation of resources as well as quality networks.

Another view to review the definition of social infrastructure could be to look at this subject through the lens of geography, urbanism, and sociology. Abundant economic opportunities, cultural amenities, and engaging architecture could be some aspects to describe a good region, city, neighbourhood, or area. However, public character should not be neglected as a key feature of being a good region for people to live in (Latham and Layton 2019). For an individual to be a part of a community, it would be necessary to access the places to make connections so that a bond between the people of the community happens and social networks shape by which important resources in times of stress are being provided. In this regard, a prominent sociologist defines the social infrastructure as all the infrastructures that participate in the development and maintenance of the connections in the society and he expresses that these infrastructures could include both physical and institutional infrastructures (Klinenberg 2018). Libraries, schools, gyms, stadiums, universities, and communal centres are examples of places that can play a critical role in the social life of a region. Social infrastructures can also be characterised as places in which not only the instrumental desires are met but also can provide a sense of welcoming and inclusion (Latham and Layton 2019). In other words, as it is stated by Klinenberg (2018), social infrastructures are responsible for promoting public life as well as preventing, social isolation and segregation, and eliminating discrimination. He even went further and inclusively defined the social infrastructure as all public institutions such as libraries, schools, gyms, public baths, yards, gardens, communal organisations, churches, mosques, and all other established physical spaces that help people to socialise and added that also commercial establishment such as scheduled markets could be an important part of the social infrastructure family (Klinenberg 2018).

Both physical and social infrastructures are necessary for the development, of the economy, and society. In this regard, implications of the physical and social infrastructures have always been of great importance for many researchers. For instance, Chong and his co-workers (2007) reported that the welfare in a community with access to the set of physical infrastructure is better than those without access to those infrastructures. Metwally and her co-workers (2007) describe the physical infrastructure as a footing for the social infrastructure to work properly and deliver. It is reported that social infrastructures also could enhance the economic growth and social development of a nation's community (UN.ESCAP 2006). In a study by Leipziger, it is concluded that if physical infrastructure becomes more available in poor or developing countries education and therefore income and welfare will be improved (Leipziger 2003). It is asserted in the literature that investment in social infrastructure as well as physical infrastructure could reduce social inequity as well as increase social development (Calderón and Servén 2008). For instance, the productivity of business goes hand in hand with the increase in electricity infrastructures due to a decrease in power outages and surges (Gnade, Blaauw et al. 2017). Human capital is a key dimension to assessing the development of a country (Kumari and Kumar Sharma 2017) and the development and the increase of social infrastructures is a key to the development of human capital and increase the quality of life (Tiwari 2000, Guisan 2009). Besides, the investment in social infrastructures could help decrease the skilled shortage while increasing their productivity as well as efficiency (Kumari and Kumar Sharma 2017). Therefore, this will lead to an eradication of poverty while enhancing the livelihood and socioeconomic status (Guisan 2009, Pal 2010).

A number of researches have also been conducted on the reciprocal effects of investing in physical and social infrastructures. For instance, Gnade, Blaauw, and Greyling (2017) reported that the investment in physical infrastructure could enhance literacy in different ways, for example, water and sanitation-related infrastructures could improve education through the reduction of waterborne diseases and therefore the decrease of absent students (Brenneman and Kerf 2002). Electricity production infrastructures decrease illiteracy by providing the opportunity for students to continue studying after the daylight as well as using new technological hardware (Bond 1999).

Referring to the earlier discussion of infrastructure systems being at the core of SDGs implementation within the community, Figure 1.3 shows the types of infrastructure and potential contributions to the goals of SDGs. Governance of the interconnected infrastructure system is underpinned by the necessary strategy and resources. Necessary infrastructure is then developed under the social or economic infrastructure categories for supporting the implementation of the SDG goals within the specific community. As seen, due to the virtue of the infrastructure and underlying contributions to the specific eight SDG goals namely goals 6, 7, 8, 9, 11, 12, 13, and 15, the types of infrastructure can be branded under the economic categories. Similarly, concerning the remaining nine goals namely goals 1, 2, 3, 4 5, 10, 14, 16, and 17, the underlying infrastructure systems meet the characteristics of the social infrastructure types regarding the earlier classification as in Figure 1.2.

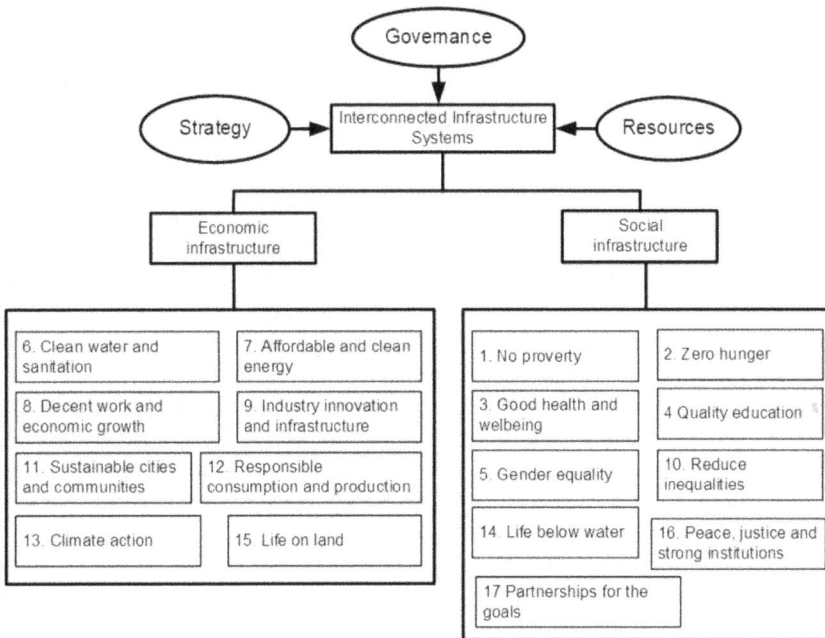

Figure 1.3 Infrastructure and development.

1.5 Rural infrastructures and their significance

While the role of rural infrastructure for supporting activities around agriculture and other associated industries, the necessity for the provisions in appropriate implementation of the SGDs is inevitable. Besides, the economic development of the rural regions as well as the quality of life in those regions could be significantly impacted by these infrastructures (Satish 2007). However, the infrastructure systems in rural contexts are prone to a low rate of returns on investment and a high incremental capital-output ratio (Satish 2007). In the context of the traditional measures of infrastructure performance, rural infrastructure is also of high risk with usually substantial investment of public funds, long construction duration and lack of maintenance in the operational periods (Satish 2007).

There are varied insights into the classification of rural infrastructure considering the underlying functions and services it offers across numerous areas. For instance, Dlugozima (2019) suggested rural infrastructure is better divided either into technical or social categories (Dlugozima 2019). Again, both technical and social infrastructure could be categorised into commercial (self-financing) and social (subsidized from the budget). Examples of technical-commercial are energy, sanitary facilities, etc. Examples of technical-social are roads, streets, zebra crossings, lighting, etc. Examples of social-commercial are entertainment, tourism, trade, gastronomy, banks, insurance, etc. Examples of social-social are education, healthcare, social service, culture, security, public order, etc.

Agricultural infrastructures could be part of rural infrastructures that have growing significance in the economy and development (Satish 2007). Agriculture-based infrastructure in rural settings is inseparable from other rural infrastructures and this is because the agricultural infrastructures alone could not lead to a holistic development of a particular rural region. For example, irrigation facilities as well as the associated machinery are purely agricultural infrastructures while the rural region cannot use them properly in the absence of good health infrastructure in that rural area. For this reason, some researchers such as Fosu et al. (1996) suggest that agricultural infra-structures need to include irrigation infrastructures, means for transportation, services for storage and preservation, facilities for processing, agricultural research and development facilities, access to water resources, access to information and communication technologies, health infrastructures, public services, commercial units, financial and funding institutions, etc. Satish (2007) also categorised agricultural infrastructures into three categories based on the intensity of capital expenditure, such as roads and bridges, services, and institutional infrastructures. He emphasised that the type, nature, and characteristics of agricultural infrastructures depend on factors such as physical-climatic, socio-cultural, and institutional (Satish 2007).

While rural infrastructure especially agricultural infrastructure is the backbone of many agrarian communities, efficiency in the operation of this infrastructure is another important consideration for reaping the benefits and supporting the growth. Ease of access to the markets for the farmers, and availability of finances are also some of the important considerations of the existence of rural infrastructures that could lead to agricultural development (Satish 2007). Based on the research conducted by the World Bank (1994), the provision of infrastructure not only could help the growth of productivity in farms but also could enhance the non-farm economy and consequently employment and therefore it potentially leads to a reduction of poverty. Such an outcome links to the SDG Goals 1 (no poverty) and 2 (zero hunger) as shown in Figure 1.3. Statistically, however, the absence of some of the ancillary infrastructure provisions such as storage facilities and access roads could lead to the loss of up to 15% of the crops and fresh produce between the farm gate and the consumer (published by the World Bank (Ayres and Mccalla 1997) (Images 1.4 and 1.5).

It is now established that the connectivity of rural regions to urban areas has a prominent role in the development of the rural regions (Singh, Batra et al. 2007, Kumari and Kumar Sharma 2017). This connectivity will not happen unless a sincere effort is made to roll out location-based and context-specific infrastructure in the rural region and support addressing the SDGs in localised settings. This will not only contribute to the development of the rural infrastructure lead with the purpose for reduction of poverty,

Image 1.4 A highly engineered vernacular weaving machine with paper-based punch cards for producing high-value traditional female attires "Mekela Sador" in Assam.

Image 1.5 Items showing manually-weaved "Mekela Sador" in Assam with a market value could be up to USD 1,500 per pair.

unemployment, and price distortions, but also it will increase the productivity in the agricultural sector, and raise the standards of living (Antle 1984, Binswanger, Khandker et al. 1993, Buys, Deichmann et al. 2006).

Along with the rural infrastructure provision, costs of operation are also an important consideration when comes to providing access to all segments of the market involved in the business processes. In the absence of rural infrastructure or due to the deterioration of these infrastructures, the cost of marketing could surge dramatically so that poor farmers are left isolated from the market (Satish 2007). Besides, the traders' ability to commute, communicate, and trade with these remote regions could be dwindled resulting in further isolation of rural farmers and increasing the poverty in these regions. The research by Jazairy and his co-workers (1992) asserted that the construction of rural roads has an inevitable role in the improvement of productivity and the increase of production in the agricultural sector as those roads could lead to better land use for cultivation due to the improvement in the accessibility both to the land and market. The research conducted by Binswanger et al. (1993), in which 13 states of India have been studied, also demonstrates the significance of rural infrastructures in plunging the costs associated with transportation, transactions, and credit services while the increase of market access, and funds. In addition to the impacts of rural infrastructures on agricultural-related activities and economy, the fact that the majority of households in the rural and regional areas have more than one income activity (Satish 2007), which could be either agricultural or non-agricultural infers that proper access to public infrastructures diversifies the livelihoods in these regions. In other words, with proper access to infra-structures, the members of a rural area could follow their preference for the job including offering helping hands for others in the agricultural activities. With this, a more diversified labour market will appear in the rural region and the households will not put all their eggs into one basket as a result of which they may not lose all their income due to natural risks or other adverse eventualities within the agricultural activities. In a study in the Tanzanian context, it is reported that non-agricultural activities have been significantly increased due to better transportation infrastructures (Lanjouw, Quizon et al. 2001). Considering 600 households in peri-urban Tanzania, Lanjouw et al. (2001) reported that access to infrastructure as well as education could affect the non-agricultural income of the households in peri-urban areas (Lanjouw, Quizon et al. 2001). The development of rural infrastructures also could lead to the access of farmers and rural products to a wider market letting them not only compete but also to access cheaper supplies (Satish 2007).

1.6   Summary

Based on a wide literature search, this chapter highlights the roles and significance of rural infrastructure from the rural community perspective. While the UN-based SDG framework is considered to be one of the key requirements for empowering rural communities, how the SDGs are linked to the relevant infrastructure is discussed in detail. The concept of SDGs being linked to the infrastructure systems is one of the key novelties for supporting agrarian communities in rural settings and the underlying contributions to their sustained growth and development. The significance of the interconnected infrastructure network and associated support provisions is also highlighted taking into consideration of the holistic and lifecycle viewpoint. The lifecycle of the infrastructure development to operations is looked at in the context of production to supply chain and retail business opportunities for potential engagement and empowerment of the rural community in its entirety.

The book is arranged across eight key chapters focusing on diverse aspects associated with planning, development and implementation of infrastructure systems in rural settings. One of the key arguments put forward by the author in this book is that knowledge, processes tools, and techniques required for rural infrastructure development are not simply an extension of the urban areas. Rather rural infrastructure development is underpinned by a myriad of intrinsic factors and context-specific considerations.

Highlighting the significance of infrastructure for smart villages, Chapter 1 has reintroduced the concept by connecting this book and its contents to its predecessor "Planning, Housing and Infra-structure for Smart Villages". It briefly discussed the types of infrastructure (hard, soft, physical, social, economic, environmental) and introduced a roadmap with a clear articulation of the UN's SDGs as an underlying principle for guiding the development and implementation of rural infrastructure in the subsequent chapters.

Chapter 2 highlighted the network of information that relates the different types and parts of infrastructure, while also hinting how infrastructure affects the other non-infrastructure elements and institutes of development. The literature in this chapter is a cursory glance at how infrastructure is dealt with in different countries at the rural and regional levels (and community level). The chapter delved into a fresh articulation of infrastructure planning principles from the bottom-up community perspective where factors such as culture, heritage, diversity and inclusions, connectivity and multifunctionality, continuity, and governance are discussed in detail.

Chapter 3 discussed the role of the economic infrastructure within a general context before contextualising it to the rural and regional context. It explained the basic form of economic infrastructure that is necessary for a functioning ecosystem, that is, the presence of agriculture, local produce, local markets, and local skill development that can contribute to the economy of the village/s. The discussion in this chapter extended to the debate of how each infrastructure type links to various facets of well-being and development.

Chapter 4 discussed the way decisions are made in the rural community in general or global contexts. It is argued that due to the different rural contexts and differences in rural economic infrastructure from urban areas as established in the previous chapters, the elements that are prioritised in the villages or within a rural community are different. As the priorities are governed by the Universal Social Protection (USP) of the community, the decision-making process is different. The literature in this chapter demonstrated how decisions are made based on the USP of the village and how it has been used by the community to facilitate development in their rural context. In this chapter, different approaches of development projects are highlighted with a focus on roles and responsibilities undertaken by the community. In conclusion, the chapter argued towards a decision-making process or a policy framework that involves community integration ensuring context-specific outcomes.

As part of developing an infrastructure framework/policy for smart villages, Chapter 5 addressed sustainability and different aspects at the rural level that influence and contribute to a sustainable infrastructure. Based on the typical definition, it looked at economic, social, and environmental sustainability and extended the discussions on how innovation and technology are introduced to society for ease and comfort, while also weighing whether these introductions are sustainable in the rural context. For example, the questions such as whether or not public transport is needed and how to operate and maintain it in a way that the rural community can afford and benefit need to be answered. Questions such as, is there a way to provide public transport infrastructure that ensures that people get where they need to go and whether the income generated from transport users suffices the operation and maintenance of the infrastructure, were discussed in a rural context.

Chapter 6 discussed the risks and vulnerabilities of introducing new technology and building necessary resilience in the rural community. It is the other side of the coin of addressing sustainable infrastructure. The literature discussed practices that can be/are incorporated to make the infrastructure resilient to potential failure in the system due to minor hitches in the process or natural disasters.

Highlighting the significance of costs, budgeting, funding, and procurement in providing infrastructure and facilitating development, Chapter 7 discussed the issues around rural investments including the factors causing failures, shortfalls, and overspending in infrastructure development. Following from the previous chapters and recognising that for sustainable infrastructure, not all the same elements are present within the rural ecology, this chapter extended the discussions on the facilities, skills, materials, and other resources that need to be procured or outsourced for smoothly functioning systems. It discussed the accessibility to a facility and the hurdles that are present in obtaining it. The literature addressed a few innovative ways that communities have managed access to basic infrastructure such as water, electricity, internet etc.

Having discussed the rural infrastructure as a network/system of functions, facilities, and skills, this final chapter discussed governance with a particular focus on stakeholders, their roles, responsibilities, and functions in facilitating development. Some of the key roles highlighted are the role of institutions at multiple levels, the role of investors, the role of practitioners, role of the community. The role of good governance was discussed from a range of perspectives such as supporting expansions, interconnections, and integration of communities and businesses for achieving sustained growth and maintaining an upward trajectory in the development process. Focusing on the effective implementation of SDGs as a catalyst for promoting sustainable livelihood and resilient community, the significance of participatory governance in rural settings is discussed with examples. Finally, the role of good governance in planning, development, and operation of purpose-built infrastructure systems and thereby capitalising on the "Rural Share" of the 40% of rural communities through smart villages are highlighted.

References

Aldrich, D. P. (2010). "The power of people: social capital's role in recovery from the 1995 Kobe earthquake." Natural Hazards **56**(3): 595–611.

Aldrich, D. P. and M. A. Meyer (2014). "Social capital and community resilience." American Behavioral Scientist **59**(2): 254–269.

Antle, J. M. (1984). "Human capital, infrastructure, and the productivity of Indian rice farmers." Journal of Development Economics **14**(1): 163–181.

Ayres, W. S. and A. F. Mccalla (1997). Rural development – from vision to action, Washington, DC, World Bank Group.

Binswanger, H. P., S. R. Khandker and M. R. Rosenzweig (1993). "How infrastructure and financial institutions affect agricultural output and investment in India." Journal of Development Economics **41**(2): 337–366.

Bond, P. (1999). "Basic infrastructure for socio-economic development, environmental protection and geographical desegregation: South Africa's unmet challenge." Geoforum **30**(1): 43–59.

Brenneman, A. and M. Kerf (2002). Infrastructure & poverty linkages. A Literature Review, Washington, DC, The World Bank.

Brown, A. R. (2021). "'Driving down a road and not knowing where you're at': Navigating the loss of physical and social infrastructure after the Camp Fire*." Rural Sociology **87**(1): 3–25.

Buys, P., U. Deichmann and D. Wheeler (2006). Road network upgrading and overland trade expansion in sub-Saharan Africa, Development Research Group World Bank.

Calderón, C. and L. Servén (2008). Infrastructure and Economic Development in Sub-Saharan Africa, World Bank Policy Research Working Paper No. 4712.

Chirisa, I. and V. Nel (2021). "Resilience and climate change in rural areas: a review of infrastructure policies across global regions." Sustainable and Resilient Infrastructure **7**(5): 380–390.

Chong, A., J. Hentschel and J. Saavedra (2007). "Bundling of basic public services and household welfare in developing countries: an empirical exploration for the case of Peru." Oxford Development Studies **35**(3): 329–346.

Długozima, A. (2019). "Social infrastructure of burial nature in the spatial development of rural areas in Poland." Infrastructure and Ecology of Rural Areas **19**(1): 79–94.

Evans, B. and M. O'Brien (2015). Local governance and soft infrastructure for sustainability and resilience. In: Fra.Paleo, U. (eds), Risk governance: The articulation of hazard, politics and ecology. Dordrecht, Springer Netherlands: 77–97.

Flora, C. B. and J. L. Flora (1993). "Entrepreneurial social infrastructure: A necessary ingredient." The Annals of the American Academy of Political and Social Science **529**(1): 48–58.

Fosu, K., N. Heerink, E. Ilboudo, M. Kuiper and A. Kuyvenhoven (1996). Effects of public goods and services on food security in Ghana and Burkina Faso: theory and modelling approaches. Proc. Int. Conf. on Sustainable food security in Central West Africa, G. Benneh et al. (eds.). Reseau SADAOC.

Fourie, J. (2006). "Economic infrastructure: A review of definitions, theory and empirics." South African Journal of Economics **74**(3): 530–556.

Gibson, J. R. (2017). Built to last: Challenges and opportunities for climate-smart infrastructure in California, USA, Union of Concerned Scientists.

Gnade, H., P. F. Blaauw and T. Greyling (2017). "The impact of basic and social infrastructure investment on South African economic growth and development." Development Southern Africa **34**(3): 347–364.

Guisan, M.-C. (2009). "Indicators of social well-being, education, genre equality and world development: analysis of 132 countries, 2000–2008." International Journal of Applied Econometrics and Quantitative Studies **6**(2): 5–30.

Hirschman, A. O. (1958). The strategy of economic development, New Haven, Yale University Press.

Jazairy, I., M. Alamgir and T. Panuccio (1992). The state of world rural poverty: an inquiry into its causes and consequences, New York, New York University Press for the International Fund for Agricultural Development.

Kavanagh, S., A. Shiell, P. Hawe and K. Garvey (2020). "Resources, relationships, and systems thinking should inform the way community health promotion is funded." Critical Public Health **32**(3): 273–282.

Klinenberg, E. (2018). Palaces for the people: How social infrastructure can help fight inequality, polarization, and the decline of civic life, Crown.

Kumari, A. and A. Kumar Sharma (2017). "Infrastructure financing and development: A bibliometric review." International Journal of Critical Infrastructure Protection **16**: 49–65.

Lanjouw, P., J. Quizon and R. Sparrow (2001). "Non-agricultural earnings in peri-urban areas of Tanzania: evidence from household survey data." Food Policy **26**(4): 385–403.

Latham, A. and J. Layton (2019). "Social infrastructure and the public life of cities: Studying urban sociality and public spaces." Geography Compass **13**(7): 1–15.

Leipziger, M. D., M. Fay, W. Quentin and T. Yepes (2003). Achieving the Millennium Development Goals: The Role of Infrastructure, November (Source: CiteSeer).

Metwally, A. M., A. Saad, N. A. Ibrahim, H. M. Emam and L. A. El-Etreby (2007). "Monitoring progress of the role of integration of environmental health education with water and sanitation services in changing community behaviours." International Journal of Environmental Health Research **17**(1): 61–74.

Pal, S. (2010). "Public infrastructure, location of private schools and primary school attainment in an emerging economy." Economics of Education Review **29**(5): 783–794.

Prud'homme, R. (2004). Infrastructure and development, World Bank Washington, DC.

Rietveld, P. and F. Bruinsma (1996). Is transport infrastructure effective? Transport infrastructure and accessibility: impacts on the space economy, Netherland, Springer Berlin, Heidelberg.

Satish, P. (2007). "Rural infrastructure and growth: an overview." Indian Journal of Agricultural Economics **62**(1): 32–51.

Singh, S., G. S. Batra and G. Singh (2007). "Role of infrastructure services on the economic development of India." Management and Labour Studies **32**(3): 347–359.

Summers, D. (2003). Longman dictionary of contemporary English, Longman.

Tiwari, A. K. (2000). Infrastructure and economic development in Himachal Pradesh, Indus Publishing.

UN.ESCAP (2006). Enhancing regional cooperation in infrastructure development: including that related to disaster management., ECONOMIC AND SOCIAL COMMISSION FOR ASIA AND THE PACIFIC, UNITED NATIONS.

United Nations (2015). The UN Sustainable Development Goals. United Nations, New York. Available at: http://www.un.org/sustainabledevelopment/summit/ (accessed 10 January 2024).

World Bank (1994). World Development Report 1994: infrastructure for development. World Development Report 1994, New York, Oxford University Press.

2 Infrastructure planning

Contexts and principles

2.1 Introduction

Infrastructure planning in general terms has a history of being customary reflecting the traditions and characteristics within the societies. Infrastructure planning can incorporate the diversities within the considerations such as ancient rules of land ownership, inheritance of lands, and rights of individuals and communities, etc. Planning in rural and regional contexts requires both protection of the inheritance as well as development of the physical, environmental, and social capital. Infrastructure planning needs to balance the economic goals by conserving natural resources while facilitating the access of rural people to services, housing, and social and physical infrastructures. Planning in the rural region should be such that the results not only bring about the burgeoning of the economy with purpose-built facilities but also lead to fairness, social inclusion, as well as poverty reduction by imparting consistent and sustained services and accessibility.

After the Second World War, philosophies of rural planning and development for countries such as the United States, China, and India were far different for decades as the role of rural areas for the development planning of each of these countries was different. The idea of modern farms, which pioneered in the United States, was far more different from other countries. While modern farms in the United States were established in collaboration with educational institutions and government with the idea of increased and sophisticated production with higher efficiencies, the priorities and ideologies were much different across other countries. For instance, modern farm concept with a target of serving national and international markets. However, concerns such as social equity and social justice especially in third-world countries and uncertainties about the redistribution mechanisms of productive rural assets and their incomes have led to a new definition of planning in rural and regional contexts (Dandekar 2015).

Although in some countries the term "rural planning" is not explicitly present in governmental policies and plans, every enactment concerning agriculture, communication, transport, welfare, education, etc., could extensively affect the rural areas as they may change the configuration and physical

DOI: 10.1201/9781032622323-2

features of the rural regions. They also could impact the available jobs and change the socio-demographic features of a region as well as the access to services and infrastructures. However, there are some instances in which the so-called sectoral policies and aims have been applied to rural areas. These cases are often trying to address inequity and social justice issues in rural planning (Dandekar 2015). Rural planning needs encompass both explicit and implicit plans that could affect the rural and regional economy, society, and environment.

In the 19th century, the role of agriculture and agrarian society was quite central to economic growth and rapid development including in the United States. During the second half of the 20th century, population and independence and the consequent task of nation-building in the heavily populated countries of China and India which predominantly were agrarian societies, agricultural economy became a priority. However, over the years political games and governmental interventions have affected the rural planning approaches across these countries which eventually shifted the country-specific priorities with prevailing controlled and market-driven strategies as opposed to common goods and well-being (Dandekar 2015).

The three countries, namely the United States, China, and India, have chosen, as examples, of states representing developed policies concerning rural planning which were vastly different from the underlying philosophies of the role of rural and regional districts in a national development context. While the United States' planning was based on the free market and the market-led orientation of the "first world" nations, China's so-called second-world planning policies were more based on community-led forces. China's first adopted policies originated from the former Soviet Union's rural-cooperative-based model of rural planning. However, over time China's planning was more inclined towards a market-based industrialised approach which has eventually proven to be the best in their rapid urbanisation process. Contrasting to the views of the United States and China, India being a third-world economy at the time sought an intermediate approach of developing the rural infrastructure based on a strong agrarian viewpoint.

Despite a scattered and country-specific viewpoint in infrastructure planning, underlying principles of an agricultural-based economy played a major role in rolling out infrastructure as a national priority across many countries. In general, it can be concluded that the agricultural sector has always had a far-reaching impact on rural development across many nations including the United States (Dandekar 2015).

Over time especially in the late 19th century with the global phenomenon of the free global economy, many countries including the United States infrastructure planning have taken a significant shift to developing agricultural infrastructure with a modern and urban-centric viewpoint. Subsequently, rural planning is intertwined with economic sectors such as services, manufacturing, and infrastructure. The emerging economy coupled with the necessity for integration of rapid evolution of information technology (IT) resulted in more centralised development within the metropolitan regions. This centralisation eventually affected rural and regional planning along with the effect on rural communities about accessibility, public health, education, and social welfare services compared to the metropolitan areas.

While over time a shift in rural planning policies became apparent with the need for integration with urban infrastructure across many countries, due to uneven population spread and lack of sufficient population density in rural and regional areas, such a policy was found to be not quite effective. Considering the economy of the scale and all-round economic development, efforts have been made to create zones for rural initiatives to potentially attract entrepreneurs including support for career training and development programs (Dandekar 2015). However, these efforts are too negligible for sustained growth of rural and regional developments with necessary infrastructure provisions. With the advent of advancement in telecommunication and high-speed Internet bandwidth, the focus of planning efforts in rural and regional contexts has re-emerged for being centred around the remote accessibility and virtual connectivity between rural, regional, and urban areas. The reason behind these foci is that the connectivity of the dispersed population through high-speed internet connections or infrastructures is prioritised over the physical infrastructure which eventually voided the need for separate planning policies between rural and urban infrastructure considerations (Dandekar 2015).

After the Second World War and the independence of numerous countries from the so-called colonial rules, some countries including China adopted Soviet-style rural planning inferring a central powerful government. However, another group of countries with low economic status have adopted a much more centralised but hierarchical planning approach with a target of modernising the whole economy through agriculture-based industrialisation incorporating both public and private agencies (Dandekar 2015). In such models, rural planning plays a central role in national economic development, and not only does it aim to boost production, but also it has a key role in the redistribution of wealth endorsing social justice and human well-being. Due to the virtue of vast majority of the population in those countries being rural and regional, addressing social justice and human welfare in government planning and operations was a need of the time (Dandekar 2015). While social equity, resource sharing, and societal well-being through agriculture-based industrialisation and income were at the core of planning and development decisions among the government across many agrarian countries in the 19th century, in recent years, those priorities have been shifted towards more on manufacturing-based industrialisation and urban-centric development (Dandekar 2015). The agriculture-based planning models have been replaced with integrated development models for wider distribution of resources and globalised mobilisation of goods and services.

Referring to the rural-centric Indian economy and planning priorities encompassing social justice and equity, Mosher (1976) summarised planning for the rural and regional development into four key groups as depicted in Figure 2.1 (Image 2.1).

Figure 2.1 Characteristics of planning and rural development.

Image 2.1 A tourist lodge in the middle of a village in Assam for experiencing nature-based living and supporting non-agricultural economy.

Although Mosher (1976) pioneered the conceptual underpinnings of rural planning through these key groups, the principles were vastly applicable in the international context. Transformation of society and morals, justice in society, economy and distribution, and the participation of the community in decision-making are still objectives of rural planning approaches in many countries though the policies, capacities of implementation, and emphases could be different.

2.2 Green revolution and sustainable agriculture

While corporate farming with high-yield mechanised agricultural production supported a green revolution across many countries after the Second World War, with the emergence of the side effects of intensified agricultural cultivation and its negative impacts on the environment, sustainable agriculture and its relevant policies have become a new way of perceiving agriculture and rural planning at large (Cochrane 2003). The idea of sustainability in agriculture, forestry, mining, etc. encompassed nature-based rules without over-exploitation of natural resources but with a significant focus on the conservation of nature. These concerns have led to the recruitment of new approaches in which traditional sustainable practices have been recognised. These new approaches to planning for rural areas try to implement scientific solutions as well as following the traditional self-sustaining forms of agriculture (Pichón, Uquillas et al. 1999) to subdue the deteriorative effect of exhaustive agricultural activities. Accordingly, the idea of sustainable agriculture has led to a new stream of rural planning approaches based on traditional self-sustained agriculture (National Research Council 1989).

With growing awareness of sustainable agriculture and underlying conservation principles, the planning of rural infrastructure came under scrutiny where incorporating nature-based rules to maintain rural characteristics became a norm (Dandekar 2015). Subsequently, the development and support of agricultural and in a more general aspect, natural resources, conservation, and protection of these resources as well as pillars of sustainability have become the emerging principles for supporting the rural planning and development (Dandekar 2015). Thus finding a way to preserve the population in rural areas while resurrecting or establishing the local economy became a challenge in contemporary rural planning, especially in the 20th century practices. In this regard, some scientists have come up with the idea of amenity-led rural planning for places with a particular emphasis on the characters such as picturesque views, scenic routes, coastal regions, historic sites, ancient buildings, etc. (Marcouiller, Clendenning et al. 2002). Due to the potential for increased migration and tourism activities, this type of planning, however, may result in higher living costs and unaffordable amenities for the local population and farm workers. This may eventually make the local workers move out to the remote areas for cheaper accommodation and living expenses. However, such a phenomenon would result in increased costs for longer commute time and contribute to deterioration in the quality of life and equity in social settings (Nelson and Nelson 2011).

Some planners propose the need for establishing manufacturing, agricultural products, and byproducts processing units in rural and regional districts as a result of which job opportunities will be created for low-salaried workers and promote higher income generation opportunities (Dandekar 2015). Adopting this principle, the United States Department of Agriculture established a Rural

Development program in 1994 to support non-agricultural housing and associated services, water and wastewater management, energy production and distribution, etc. integrating rural and regional areas (USDA 2004).

In the 21st century, the emphasis on context-specific rural policies and planning is considered to be pivotal for empowering human capital. Context-specific social infrastructure systems with localised planning priorities are at the core of enhancing the education of the community and improving the connectivity and accessibility of rural residents. However, integration of rural infrastructure with urban networks and providing seamless accessibility to respective amenities for both urban and rural populations is one of the key requirements supporting the globalisation of trades and services and promoting equable development of rural communities. Acknowledging the need for such practice, Marshal (2001) asserted the significance of the co-existence of both first and third-world economies in creating a seamless integration of urban and rural infrastructure for providing equitable access for rural communities and reducing the gaps between the rural and urban communities. The ever-increasing accessibility to the telecommunication and Internet network along with other cutting-edge smart infrastructure provisions are paving the way through a systematic similarity around the world as a result of which the ideas and approaches for rural planning could be cross-fertilised. On the other hand, indeed, the fast-paced transformation of society is engulfing developing countries and pushing towards unprecedented economic liberalisation. This may, however, result in rural planning being less prioritised over urban planning. Such a trend eventually results in the environment and justice being compromised over the economic transformations in rural and regional areas. In this regard, concerns about sustainability, and resilience of settlements and communities should be addressed in rural planning (Image 2.2).

Image 2.2 A traditional wooden boat for transporting passengers and goods including cars to Majuli Island from Nitami Ghat in Jorhat, Assam.

2.3 Principles of planning infrastructures in rural and regional context

2.3.1 Nature-based planning

Nature and the underlying ecosystem generally have a great ability to inspire planning and services in our society. This is more so in rural settings where uniqueness in the ecosystem is quite intact. The International Union for Conservation of Nature (IUCN) (Cohen-Shacham, Walters et al. 2016) defines the general term of nature-based solutions as "Actions to protect, sustainably manage and restore natural or modified ecosystems that address societal challenges effectively and adaptively, simultaneously providing human well-being and biodiversity benefits". The supporting role of ecosystems in human welfare was not quite recognised in modern scientific literature until the 1970s though this has always been part of many indigenous people's beliefs that ecosystems are at the core of servicing community in our society (Cohen-Shacham, Walters et al. 2016). However, over time, it became apparent that understanding nature and its reciprocal relationship with the community is one of the core requirements of nature-based rural planning and development (Cohen-Shacham, Walters et al. 2016).

IUCN has always been developing and trying to apply the concept of nature-based planning and some other institutions and organisations such as the European Commission have followed suit (Cohen-Shacham, Walters et al. 2016). As suggested by IUCN, Figure 2.2 highlights the eight principles for nature-based planning or solutions.

As seen, the first principle of natural conservation follows the norms of nature conservation and preservation while planning the rural infrastructure, especially for agriculture-based economies. The second principle requires seamless integration of natural conservations while addressing the challenges

Figure 2.2 Eight principles for nature-based planning.

in engineering, technological, and societal aspects. Referring to the third principle, the socio-cultural along with environmental and natural character-istics should be at the core of nature-based solutions in rural planning. As the fourth principle depicts, fairness and equity are the key requirements for promoting transparent planning and governance as well as encouraging the community's broad participation in nature-based planning. As suggested in the fifth principle, attention to indigenous knowledge and relevant consider-ation of cultural and biological diversity is of high importance in promoting nature-based planning. Thus, proactive plans for maintaining and preserving the ecosystem are highly significant for achieving support from the broader community in location-based infrastructure planning and development. As suggested in the IUCN framework (Cohen-Shacham, Walters et al. 2016), the sixth principle of nature-based planning is about the assimilation of the natural landscape at both neighbourhood, district, and state levels. This principle is particularly useful for promoting biomimicry in rural planning. The seventh principle of this method of planning is to find a trade-off by comparing the cost benefits between producing a few instant economic advantages for the development and the wide variety of ecosystem services that are going to be achievable. The eighth principle of nature-based planning emphasises the role of these solutions on the overall design of policies, assessments, and actions and stresses that these solutions should be the central part of the design to address a specific challenge. It is also important to note that nature-based solutions and planning are not a replacement for the conservation of nature they should embrace the principles of nature conservation (Cohen-Shacham, Walters et al. 2016).

2.3.2   Green infrastructure planning

The concept of green infrastructure goes back to the 19th century when the famous landscape designer, Frederick Law Olmsted, started with the notion of linking parks and neighbourhoods, which was later widely recognised as the first inspiration for green infrastructure (Benedict and McMahon 2002). In the mid-19th century, rapid industrialisation resulted in a significant influx of population in the large American cities. As a result of this influx and due to the lack of proper infrastructure in cities, the quality of life deteriorated including the spread of deceases was ubiquitous (de Oliveira and Thompson 2015). However, prioritising the infrastructure, Europe at the same time undertook some serious social reforms for establishing public parks as a way of improving the health and well-being of the working class (Ward Thompson 2011). For instance, Victoria Park in London and Birkenhead Park near Liverpool were individual parks that were open to the public and make a good example of the green interventions. These interventions have later influenced many countries and experts to recreate the infrastructure with stronger integration of green and open parks (Szczygiel and Hewitt 2000). Green infrastructures essentially have the potential to enhance public lives by facilitating physical activity, reducing stress, increasing social cohesion, and improving air quality (de Oliveira and Thompson 2015).

Green infrastructures have important roles to play in addressing environmental and social challenges. In this regard planning for such infrastructures would be in line with the objectives of both rural and sustainable development as well as sustainable agriculture (Filep-Kovács, Sallay et al. 2016). Although grey infrastructures are deemed to impact economic, social, and environmental conditions they are prone to be deteriorative for the environment if the planning does not follow the appropriate rules to achieve sustainable development goals. On the other hand, green infrastructures are networks of integrated natural or semi-natural areas that should be planned to deliver a variety of advantages to humans (Naumann, Davis et al. 2011). Along with the relatively lower financial investment, and maintenance costs in green infrastructure than for grey infrastructures (Filep-Kovács, Sallay et al. 2016), green infrastructures can support multiple functions (Ely and Pitman 2014).

Due to the virtue of green infrastructure being multifunctional which also can preserve and signify the local features in rural and local settings, these are considered important tools for rural development. Apart from the role of green infrastructure in the protection, preservation, and enhancement of biodiversity and nature in rural and regional areas (Silva 2019), they are a viable option from a sustainability perspective (Villa 2020). Many case studies support the positive effects of green infrastructures (Quintero 2012, Molla 2015, Bottalico, Chirici et al. 2016). Green infrastructure can be an important tool to mitigate climate change and increase the level of physical activity as well as enhancing the air quality. For successful green infrastructure planning, some principles need to be followed. In this regard, Monteiro and her co-workers (2020) have conducted a comprehensive literature review on 104 key references and reported eight principles being the key consideration while planning green infrastructures as shown in Figure 2.3. The principles are discussed below.

2.3.2.1 *Connectivity*

To maintain the values and services of natural systems, as well as to sustain species and diversity, connectivity is crucial. Small parks and urban forests can't sustain diverse wildlife and vegetation on their own; however, connectivity with larger urban areas and possibly expansion to semi-urban areas permits the movement of certain species, the dispersal of seeds, or even the repopulation of some patches in heterogeneous landscapes as a consequence of migration. Additionally, connectivity also serves as a transit and recreation corridor for humans, contributing to the stability of the ecosystem including a variety of services, as well as connecting different landscapes. Likewise, the goal of connectivity is to create a network of integrated green spaces that can benefit both humans and other species in the long run.

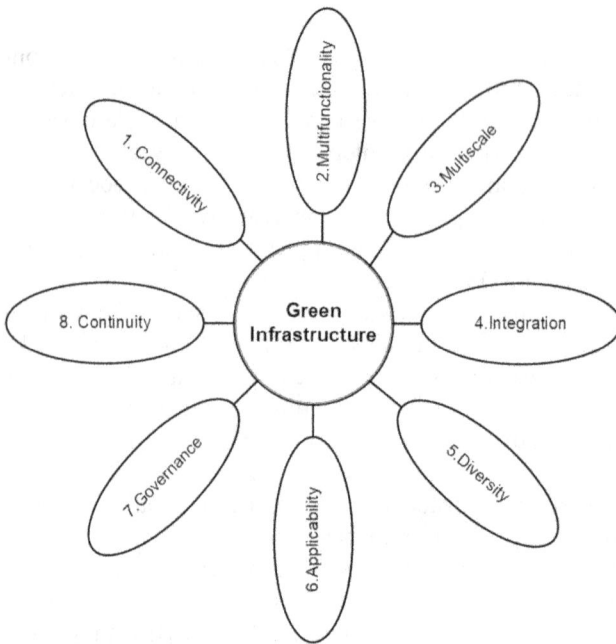

Figure 2.3 Eight principles for green infrastructure planning.

2.3.2.2   *Multifunctionality*

By way of linking the green infrastructure with several ecosystem services, including provision, regulation, support, and culture, multifunctionality of the infrastructure system can be achieved. Multifunctionality is one of the key underlying characteristics of the green infrastructure system, for performing numerous social, ecological, and economic roles and providing necessary resilience for the community. In metropolitan settings with limited space, multifunctionality not only encourages different uses and develops synergies within green spaces, but also improves the spatial efficacy of these locations. In summary, multifunctionality is an essential function of realising several interconnected SDGs in the community.

2.3.2.3   *Multiscale*

By virtue of green infrastructure being multifunctional, the interconnected infrastructure network intervenes at multiple different levels and scales. For instance, the construction process of a green infrastructure unit by integrating the built environment with the natural environment may require the engagement of expertise at multiple levels. Incorporation of the underlying principles of landscape interactions and bigger natural reserves within the built form may require both local and regional perspectives which essentially demands integrated input across numerous levels and scales. Fortunately, due to the flexibility and adaptability nature of green infrastructure, such an integrated practice is possible in the development process.

2.3.2.4 Integration

The interaction and connections of green infrastructures as well as other urban structures such as "grey infrastructures" are the core aspects of promoting integration. The concept of integration not only depicts the underlying relationships between the green and grey infrastructures but also represents the interactions of the natural landscape with the built environment. Typically, these ideas are closely applied in building designs where numerous different infrastructure types are required to be integrated to make the building functional and fit for the purpose.

2.3.2.5 Diversity

Green infrastructure places significant emphasis on the number, quality, and variety of urban green areas as well as the approaches used to address various problems within the community. Numerous types of nature-based approaches may be used to solve a variety of different problems in both rural and urban areas. For instance, the natural drainage of stormwater from buildings may require diversified approaches and processes for creating both engineered solutions and aesthetic considerations within the infrastructure systems. These solutions can have a more controlled or natural approach, and they can be used to a greater or lesser level. Similarly, various technical considerations within the blue infrastructure system while developing green infrastructure required diversity in materials, sizes and shapes, underlying provisions and workmanship, etc.

2.3.2.6 Applicability

Following the trend and due to increased environmental consciousness, many public and private organisations create green infrastructure plans which typically require significant capital expenditures in nature-based approaches especially in the urban areas. However, from an economic viability point of view, such investment may not be always feasible in the first place due to a lack of demand or interest. But over time with the change of demography, affordability, demands, and interest, the facilities may need repurposing from usability, functionality, and viability perspectives and these requirements should be considered while planning and designing green infrastructure systems. In other words, green infrastructure planning must take into account the projects' applicability, adaptability, and implementation. This determines whether the plan is feasible in the long run, can be developed in entirety or a phase-based manner and whether the solutions presented can be tailored to the area under consideration.

2.3.2.7 *Governance*

The goal of governance is to encourage community cooperation with the actors of government in the processes of planning. Since green spaces provide a variety of leisure activities that are centred on people and their administration and upkeep are directly dependent on the populace, the concept the cooperative governance is highly significant in the construction and adoption of green infrastructure. The plans and design of the green infrastructure systems and their integration with other types of infrastructure for functional effectiveness need to be discussed for public discourse before being finalised and implemented. Otherwise, the success of green infrastructure will not be acknowledged and supported by the local people, preventing its aims and goals from being achieved. This will happen if people do not feel they are part of the planning process.

2.3.2.8 *Continuity*

The absence of post-implementation supervision or realistic measures of the results of the ecosystem services and activities that green infrastructure projects promise to deliver has been a fundamental problem in these initiatives. In this view, frequent management, investment, and updates are necessary for green infrastructure to be adequate, and governments must be able to regularly disseminate new information on existing projects, aims, accomplishments, and plans for green and blue areas. In this regard, well-defined monitoring systems or reports detailing the development of the intended green projects are required for the continuity planning of the green infrastructure systems.

2.3.3 *Grey infrastructure planning*

Within a built-up civilisation, grey infrastructures are highly essential for facilitating goods and services, supporting businesses and organisations, and offering the necessary support for carrying out the variety of core functions within the broader infrastructure systems (Spatari, Yu et al. 2011, Harrison, Bobbins et al. 2014). The functioning and effectiveness of grey infrastructure, such as roads, sewers, water, drainage, power lines, etc. rely on the broad network of the entire infrastructure systems. As depicted in EEA (2015), grey infrastructure serves as a shaping mechanism for urban patterns such as "compactness, sprawl, urban form, urban design, interconnection of streets including the mobility patterns, density of the inhabitants within the geographical jurisdictions" (e.g. housing, connectivity, mobility, accessibility). The majority of today's grey infrastructure development is managed by the people and for the people the values of which are recognised and appreciated for the functioning of the society across the settlements (Joubert 2018).

Grey infrastructure, in general, frequently deals with urban commons. The atmosphere, woods, rivers, fisheries, or grazing land are historically considered "urban green commons" (Joubert 2018). However, one of the key and emerging notions of such commons is being intrinsically linked to public goods, such as open spaces, markets, free public health care, and infrastructure that enables social and economic viability (Cilliers, Cilliers et al. 2012). According to Barthel et al. (2014), the principles of urban green commons usually do not quite support land privatisation because it reduces human interaction with nature and weakens ecological practices and functions. For sustainable development to be fully included in all urban commons, grey infrastructure's impact on ecosystem-based infrastructure and vice versa must be recognised (Roe and Mell 2013).

Grey infrastructures are always in interaction with ecosystem-based infrastructures. Along with the increasing trend of globalisation, the interaction spans of the grey infrastructure systems also increase across multiple land uses in the form of tapping aquifers, rainwater drains, roads, railways, airports, sea wharves and ports, and sewage networks. Infrastructure development alters the natural environment in settlements, resulting in a change of spatial patterns in rural to urban regions. Among the suite of individual infrastructure, roadways, for example, usually affect the spatial patterns across larger dimensions of ecosystems. This may include waterways and reservoirs, flora and fauna, and other associated land-based systems such as power, communications, and information-communication networks. In this regard, the grey infrastructures can change spatial sustainability and destabilise the demand for humans and the supply of natural resources.

2.3.4 Social infrastructure planning

The term "social infrastructure" refers to locations and spaces of public, human, or social character that are needed, first by certain parts of the community, then increasingly as the society expands. Both official and informal locations and areas that enable access to communal activities and services are provided by the social infrastructure. Both "physical" infrastructure (community buildings and public open spaces) and "soft" infrastructure collectively contribute to the social infrastructure systems that are crucial for the effective functioning of society (support services, technology, telecommunication and communication, management systems, data, and insights).

For communities to grow strong and diverse, social infrastructure is important. It has the potential to bring together diverse groups of people, nurturing social integration and promoting inclusivity and harmony in the societal context. A community may suffer if its social infrastructure is compromised. Therefore protection and safeguarding social infrastructure are fundamentally important for honouring the rights of the common public in the community. For example, any development plan that may result in relocation or closure of a community facility may affect the mental health and social well-being of the broader community. In such a situation, alternative options should be considered for safeguarding the community facility in the interest of the common public. the government should consider classifying a facility as an Asset of Community Value (ACV) if it is proposed by the local community to further protect against the loss of social infrastructure that a local community or organisation values.

2.3.5   *Rural social infrastructure planning*

Rural communities in rural areas are increasingly becoming important due to their intimacy with nature and a relatively low-cost low-maintenance lifestyle. Rural districts which are at the core of service delivery are now viewed as complex regional socioeconomic systems whose growth needs to be controlled using an integrated strategy while anticipating the establishment of suitable conditions for habitation, employment, and raising future generations. In rural regions with high social infrastructure, proper planning, integration and maintenance of the infrastructure systems are important considerations for the seamless flow of goods and services ensuring equitable accessibility across all sections of the society (Atkociuniene 2018). The infrastructure systems need to be able to entice local and social enterprises to venture on opportunities, growths, and well-being of the community.

The principal factor inhibiting entrepreneurial initiatives from developing the service industry in rural regions might be attributed to the weak local market. Consequently, compared to residents in metropolitan areas, rural residents have less favourable situations. Traditional rural social infrastructure development approaches were unable to create specific prospects for the viability, vitality, and stability of rural communities. Nature-based approaches focusing on biological biodiversity (ecological engineering) have the potential to offer effective planning solutions that are energy efficient and sensitive to cultural preservations in a sustained and durable manner (Atkociuniene 2018).

The development of the rural social infrastructure underpins the creation of sustainable societies through territorial development and the integration of rural and urban regions. This then enables community engagement activities, resource allocation, and enhancement of society groups and enterprises and potentially influences adaptive and dominant behaviour commensuration with the culture, heritage and value of the community. The development of the social infrastructure in rural areas does not happen by accident. To properly implement stated objectives, it is necessary to make specific changes to the planning, governance and management processes (Atkociuniene 2018). To ensure rural residents' social development, rural social infrastructure management includes planning for social infrastructure facilities, services, and employment size within a specific territory, coordinating capacities and development, balancing the supply and demand for services, providing opportunities to meet social needs, sustainable development of the territory, and involving residents. Social infrastructure guarantees the fulfilment of social needs and promotes the expansion of the rural, provincial, and national economies. For instance, in a study by Osumgborogwu (2016) on the effect of rural social infrastructure on the income level of the rural community of Imo state, Nigeria, a positive robust relationship has been reported between the two. In addition, improving social infrastructure gives locals a chance to pursue higher education and develop professional skills they may use in the workforce. The development of social infrastructure, particularly in rural residential areas, predetermines life quality, which is linked not only to the fulfilment of fundamental social requirements but also to self-actualisation in both work and private life. For instance, in a study by Sharp and his co-workers (2002), correlations between aspects of social structure within communities and the presence of two divergent forms of economic development—self-development and industrial recruitment—in rural areas were investigated. Considering a hypothetical viewpoint that social infrastructure is more strongly associated with the presence of self-development than industry recruiting using data gathered in a statewide sample of 99 Iowa villages, the study asserted that the availability of social infrastructure is strongly correlated with self-development. Based on their study, while social infrastructure and industrial hiring may have a reasonable association, when it comes to growth and empowerment, the community's social structure plays a major role (Sharp, Agnitsch et al. 2002). High living standards involve several presumptions, including but not limited to monetary security, physical health, educational attainment, and social integration. The circumstances for developing rural infrastructure, however, are not necessarily the on such criteria but more on the social empowerment and growth of the community at large. Without proper planning and effective implementation, however, rural communities confront the most severe shortage of basic amenities and overall social purpose.

Due to the lack of necessary social funds and also the inability to pass the viability tests of the private funds, choosing where to situate and how many of these facilities should be in a network of infrastructure is a common dilemma in social infrastructure planning for public bodies. In most cases and traditional practices, there isn't a trade-off between the expenses of delivering the service and the most outstanding possible service to the people (Bigotte and Antunes 2007). However, if planned well, social infrastructure planning issues may have a vast number of potential solutions. Usually, location-specific and need-based planning is considered highly effective in social planning but such a practice relies on community-specific demographic data and smart interpretation for understanding the context well. With the advent of the advancement of IT, digital study of demographic data in terms of data analytics and also perhaps digital modelling, context-specific planning can be aligned closely to the needs and requirements of the community. While the data-driven infrastructure planning will be further extended in the later part of the book, the detailed modelling study is not within the scope of this book, the readers may refer to the Book by Daskin (1995) or studies by Current et al. (2002), or ReVelle and Eiselt (2005) for a thorough explanation of the topic (Image 2.3).

Image 2.3   A traditional wooden boat for transporting agricultural goods and also fishing by villagers.

2.4   Contexts of planning infrastructures in rural and regional areas

With an ever-increasing trend of rapid urbanisation in the so-called urban era (Gleeson 2013, Brenner and Schmid 2014), the topic of rural infrastructure has not been receiving sufficient attention in the mainstream literature (Gkartzios, Gallent et al. 2022). Rural areas make up a large portion of the world's land (and water), including areas used to exploit natural resources, produce energy and food, agriculture, and mitigate and adapt measures to climate change such as production of solar and wind energy, etc. However, in recent years, rural areas have become a focal point for some debates due to their contentions in terms of their future growth potentials due to environmental conservation, sustainable development goals, and social justice objectives (Scott, Gallent et al. 2019). Meanwhile, due to persisting fragilities, including extreme poverty, economic inequality, and a lack of institutional capacity, many rural regions have suffered unduly compared to urban areas throughout past and current crises such as the 2008s Global Economic Crisis, Europe's Refugee Crisis, the Climate Change ongoing Crisis, and the ominous COVID-19 Pandemic (Gkartzios, Gallent et al. 2022). However, rural communities have also shown impressive resilience and, in some circumstances, a surprising capacity to transform calamities into prospects (Gkartzios and Scott 2015), demonstrating yet another aspect of the widely diverse character of rural places across the world (Marsden, Murdoch et al. 2012). This variation also applies to the politics of rural areas. While some have been linked to the rise of right-wing ideology and populist movements (Mamonova and Franquesa 2019), others have been at the forefront of neoliberal resistance movements (Shucksmith and Rønningen 2011) and progressive visions of inclusive communities (Gray, Johnson et al. 2016).

Greater focus on the rural areas, as well as the distinctions between the global and local scales, reveals a wide range of issues, politics, and planning-related difficulties, as well as the need for revitalised rural planning theory and practice that is geared towards a more varied and bold set of objectives. As noted by Lapping (2006), moving beyond the post-war concentration on agriculture, farmland preservation, and food security is an absolute necessity for rural communities to take seriously and bank on their potential for addressing some of the emerging global challenges such as climate change, global warming, etc. (Lapping 2006). In rural areas, the persistent portrayal of farming as the foundation of rural economies has led to a lacklustre and negligible planning response (Lapping and Scott 2019), even though those areas' distinctiveness and contention suggest a pressing need to reconsider the concept and practice of rural planning for a globalised world.

Rural planning (and hence, rural regions) has been given a more peripheral position in planning theory and practice, frequently focusing on constrained growth routes connected to context-specific political agendas. While there is no any single school of thought on rural planning, due to the underlying complexities around social and demographic compositions, cultural and ecological conservation, etc., myriads of planning agendas are being applied across the communities (Gkartzios, Gallent et al. 2022). In other words, while some planning objectives could be centred around preserving and protecting the agricultural and rural lands (Curry and Owen 2009), others could be focusing on natural resource exploitation to support rural-centric new industries (Tonts 2020). Some agenda is seen to be more inclined towards facilitating family-centred welfare through permissive regulatory approaches (Gallent, Shucksmith et al. 2003) while others are focused more on transforming "business friendly" rural-ecosystem with swift planning and delivery of infrastructure across rural landscapes, particularly renewable energy (Natarajan 2019).

2.5 Considerations in infrastructure planning in rural and regional contexts

Having reviewed the broad literature in the above sections, it has become apparent that urban-centric knowledge and practices may not be extended when planning the infrastructure for rural and regional contexts. There are numerous drivers and impediments associated with the community and the region at large that must be considered in the process of planning rural and regional infrastructure. Figure 2.4 summarises eight key considerations associated with infrastructure planning in rural and regional settings which are briefly discussed below.

2.5.1 *Sustainable development and policies in rural regions*

Any form of infrastructure development requires careful planning and holistic approaches to meeting the underlying intent of providing necessary support

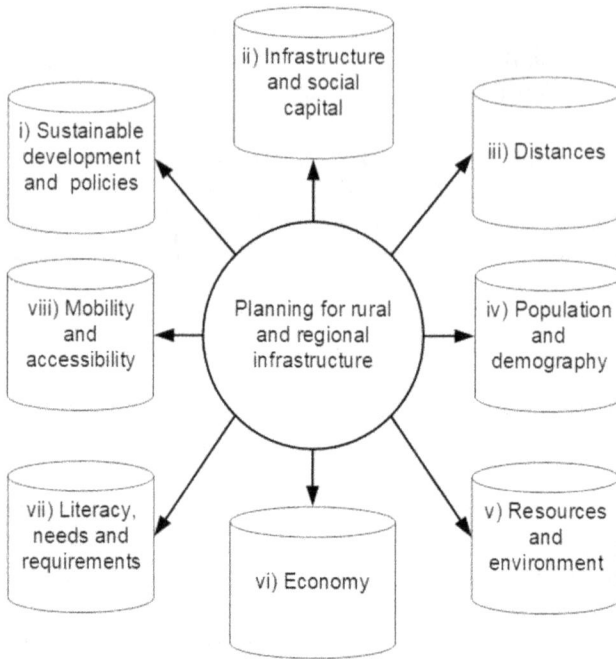

Figure 2.4 Considerations of infrastructure planning in a rural and regional setting.

and connectivity of the facilities for the community. In rural settings where nature-based integration of infrastructure systems is an integral consideration, the design and development concepts need to align the sustainable development principles commensurate with the local or regional requirements. As stated in Chapter 1, the relevant and applicable SGD goals about a particular community and their needs and concerns, potentials for development and growth trajectories provide a good basis for the development of location-specific sustainable development agenda including implementation and operations plans in rural infrastructure systems. While all 17 Goals for the UN's SDG may be considered relevant for all forms of rural infrastructure development, the application context may differ from community to community commensurate with the demographic distributions. Aligning with the SDGs, the sustainable infrastructure development agenda in rural settings must include but not be limited to issues such as poverty reduction, climate change mitigation, reduction of environmental pollution, reduction of geographical inequalities, increased prosperity, peace, and justice.

2.5.2   *Infrastructure and social capital*

Among various infrastructure types, social infrastructure has a close link with social capital due to its association with the well-being of the community at large. The associations are usually in the form of supporting the well-being of the community by improving the quality of life. Some of the key measures associated with good quality of life include conducive social space and civic engagement opportunities, connectivity and accessibility, job creation and income generation, work and life balance, environmental quality, and personal security which eventually lead to the overall improvement of the community's social and economic outcomes. According to the literature, social capital is usually a measure of a productive community resulting from exercising the above factors. The underlying measures of the productive community could be through improved trust and commitment, positive social networks and interactions including a sense of belongingness and good health and prosperity. Thus, appropriate provisions for developing and materialising these factors are essential first steps that require careful considerations and strategic decisions for effective functioning over the entire development and operational lifespans of the infrastructure systems.

2.5.3 *Distances*

The provision, accessibility, and quality of services differ from place to place and are often considered to be an indicator of the status of development of a nation. There is also a great deal of variance in such an indicator from country to country. In the less privileged nations, essential rural services are less available/accessible and of lesser quality than urban services (Potter, Binns et al. 2018). A significant body of research has focused on the urban/rural gap, suggesting that development in many nations is skewed towards cities and the remote areas are increasingly getting difficult to access due to a lack of basic infrastructure (Bates 1981, Brinkerhoff, Wetterberg et al. 2018). To reduce the urban/rural divides and connect rural communities with reasonable infrastructure access, distances between the city centre and the rural areas are important considerations for supporting integrated infrastructure networks (Allen 2010). Focusing on 17 African nations and based on 21,000 survey respondents, Brinkerhoff and his co-workers (2018) investigated the perceptions of communities about distance to metropolitan centres, access, satisfaction with services and governance. Their research asserted that geographic distances result in significant disparities in service access and satisfaction among the communities. While the urban community relies on the metropolitan authorities for their access to basic amenities, remote residents often look up to the local and national authorities to provide necessary connections and reduce geographic distances through appropriate infrastructure provisions (Brinkerhoff, Wetterberg et al. 2018).

2.5.4 *Population and demography*

Because the underlying demographic drivers differ across various population groups, rural populations do not exhibit the same patterns as urban populations (Anríquez and Stloukal 2008). The first substantial variation is fertility rates which are significantly higher in rural communities. The total fertility rate (TFR), the primary metric of fertility, is rarely estimated separately for urban and rural populations. However, when it occurs, as in Latin America, it has been demonstrated that rural women have two more children than urban women during their reproductive lives (Anríquez and Stloukal 2008). Another, more easily accessible metric, the crude birth rate, likewise reveals that the number of births per 1,000 people is more significant in rural regions (Anríquez and Stloukal 2008). The other point to be mentioned about the rural population and its role in planning is that although the ageing process is accelerating in wealthy nations, the most significant contrasts between urban and rural ageing may be observed in developing countries. Several factors contribute to the ageing process, including decreased reproduction rates and breakthroughs in medicine, diet, and technology. It symbolises significant societal success and a problem that will touch all elements of 21st-century civilisation (Anríquez and Stloukal 2008). On the other hand, for the rural areas with low population or low density, the planning should consider approaches to achieve scale economies for developing infrastructures or providing services and this would not be an easy task. For the dispersed population, the issue of connectivity stands out which not only is a threat to the quality of the services but also it threatens the growth or formation of the social capital of that community.

2.5.5   *Resources and environment*

Another significant consideration in planning rural infrastructures is the existence of resources and the quality of the environment. They could contribute to the economy of the district and lead to the attraction of entrepreneurs. As a result, job opportunities will grow, and various opportunities for entrepreneurship could happen. On the other hand, of course, these detrimental effects such consideration also should be taken into account while planning for rural infrastructure. The increase in house prices and rents in that neighbourhood and higher demand for transportation, as well as the depletion of resources, are good examples of the negative impacts of this specific action on rural planning.

2.5.6   *Economy*

A reciprocal relationship between the economy and rural infrastructure planning and community development exists. Rural infrastructure projects are often built to help the local population gain access to resources, address rural problems, and so favourably change the general rural environment (Wahid, Ahmad et al. 2017) while also assisting communities in achieving a variety of aims and objectives (Sinakou, Boeve-de Pauw et al. 2018, Tan, Shuai et al. 2019). However, the lack of these initiatives causes serious socioeconomic challenges for local populations, such as a catastrophic impact on living standards, a negative impact on quality of life, and higher spending on healthcare and education (Hussain, Zhu et al. 2017, Hussain, Maqbool et al. 2022) which results in decreased economic growth, and rising unemployment (Agarwal, Rahman et al. 2009). The infrastructure projects in the lives of the community demand functioning essential actions for community development (Hussain, Xuetong et al. 2022) in rural areas. The importance of infrastructure in sustainable rural development, such as regional economic growth via providing transportation infrastructure among areas, is one of the essential socioeconomic aspects (Li 2014). Many aspects of societal and economic activity (Dudzińska, Bacior et al. 2018) can be influenced by public infrastructure projects, including improved quality of life, gross domestic product, quality education, employment, poverty reduction, education, and improved healthcare facilities (Maqbool, Rashid et al. 2022). On the other hand, a strong economy could support planning for rural infrastructure development without compromising sustainable and equitable development. Besides, a strong economy could facilitate and encourage entrepreneurship in rural and regional areas as a result of which the social and economic infrastructures could be better built, used, and maintained.

2.5.7 *Literacy, needs, and requirements*

The literacy of the population is an important indicator for understanding the underlying needs and requirements of a community. A community with a higher literacy rate may demand more economic stimuli compared to perhaps a higher demand for agriculture-based infrastructure among the community with lower literacy rates. These considerations are important to realise or expect in the planning processes especially when it comes to the community participation spanning from the planning phase of the infrastructure to the maintenance and demolition. Besides, the literacy of the population could lead to a better understanding of the services that an infrastructure needs to provide for growth and empowerment of the specific community.

2.5.8 *Mobility and accessibility*

The notion of accessibility has sparked attention in a variety of fields during the last century, and it has been applied in a variety of contexts (Vitale Brovarone 2021). The definition of Geurs and Van Wee (2004) highlights accessibility being "the extent to which land-use and transportation systems enable (groups of) individuals to reach activities or destinations using a (combination of) transport mode(s)", and is commonly used in the area of planning. This description accurately describes how accessibility is the outcome of various interactions between land use and transportation infrastructure (Geurs, Dentinho et al. 2016, Bertolini 2017). Furthermore, accessibility is tied to capacities, which, in turn, are associated with social and cultural background, in addition to personal variables

(Vitale Brovarone 2021). Accessibility along with mobility has also been explored from various perspectives in the planning sector, both as a positive and normative notion (Levine 2020) and with a wide range of techniques, methodologies, and tools (Bertolini, Hull et al. 2019). Despite this, its comprehension and use in planning practice with integrated land-use and transportation planning still need to be expanded.

Accessibility and mobility in outlying locations have distinct aspects and dimensions from metropolitan situations and must be studied accordingly (Vitale Brovarone and Cotella 2020). Services and options are scarce and far; time-space geographies (Ellegård and Svedin 2012) domains, mobility patterns, and modal split are vastly different from those seen in populous urbanised settings. Distances are more considerable, journeys are longer, and automobile reliance is higher in these circumstances than in urban areas. Therefore, mobility is a critical component of accessibility, especially in these places where inhabitants must be more mobile to access services and opportunities (Vitale Brovarone and Cotella 2020). Spatial mobility is more than just a means of getting from one place to another; it is a "structuring feature of social interaction" (Kaufmann, Bergman et al. 2004). As a result, social inequalities grow and people who cannot drive for whatever reason are severely disadvantaged in terms of accessibility and social inclusion (Binder and Matern 2019). As a result, in rural locations, much more than in urban areas, integrated policies that operate at different levels on the many aspects of accessibility are required. At the same time, mobility is critical in determining accessibility and quality of life (Vitale Brovarone 2021) (Image 2.4).

Image 2.4 Rice farming by individual farmers on their lands and residing the village houses near the farms.

2.6 Summary

Drawing the evidence from a vast body of literature, this chapter discusses the key principles of infrastructure planning from a rural and regional context. Contrasting to the principles and philosophies of urban-centric infrastructure practices, the discussions have been extended towards nature-centric and green infrastructure as a core underpinning for the planning and development of rural infrastructure networks. Various characteristics associated with the natural and green infrastructure were discussed from both theoretical and application perspectives. Then the discussions were further extended towards the integration of grey and social infrastructure as facilitating services and contributions to the overall functioning of the community. The final section is about the core considerations in the planning and development of rural infrastructure where numerous community-specific performance indicators including SDGs were discussed and a clear roadmap for supporting planning is established.

The next chapter will look in detail at the cost and economy of the rural infrastructure systems from planning, development, operations and maintenance perspectives.

References

Agarwal, S., S. Rahman and A. Errington (2009). "Measuring the determinants of relative economic performance of rural areas." Journal of Rural Studies **25**(3): 309–321.

Allen, A. (2010). Neither rural nor urban: service delivery options that work for the peri-urban poor. Peri-urban water and sanitation services: policy, planning and method. M. Kurian and P. McCarney. Dordrecht, Springer Netherlands: 27–61.

Anríquez, G. and L. Stloukal (2008). "Rural population change in developing countries: lessons for policymaking." European View **7**(2): 309–317.

Atkociuniene, V. (2018). "The peculiarities of strategic management of rural social infrastructure development." Scientific Papers of Silesian University of Technology. Organization and Management Series. 2018. 21–33. doi: 10.29119/1641-3466.2018.128.2.

Barthel, S., J. Parker, C. Folke and J. Colding (2014). Urban gardens: pockets of social-ecological memory. Greening in the red zone: disaster, resilience and community greening. K. G. Tidball and M. E. Krasny. Dordrecht, Springer Netherlands: 145–158.

Bates, R. H. (1981). Markets and states in Tropical Africa, Berkeley and Los Angeles, University of California Press.

Benedict, M. A. and E. T. McMahon (2002). "Green infrastructure: smart conservation for the 21st century." Renewable Resources Journal **20**(3): 12–17.

Bertolini, L. (2017). Planning the mobile metropolis: Transport for people, places and the planet, Bloomsbury Publishing.

Bertolini, L., A. Hull, E. Papa, C. Silva and R. A. Ruiz (2019). Accessibility: operationalizing a concept with relevance for planners. Edited by Cecilia Silva, Nuno Pinto, Luca Bertolini, Designing accessibility instruments, Routledge: 52–81.

Bigotte, J. F. and A. P. Antunes (2007). "Social infrastructure planning: A location model and solution methods." Computer-Aided Civil and Infrastructure Engineering **22**(8): 570–583.

Binder, J. and A. Matern (2019). "Mobility and social exclusion in peripheral regions." European Planning Studies **28**(6): 1049–1067.

Bottalico, F., G. Chirici, F. Giannetti, A. De Marco, S. Nocentini, E. Paoletti, F. Salbitano, G. Sanesi, C. Serenelli and D. Travaglini (2016). "Air pollution removal by green infrastructures and urban forests in the city of Florence." Agriculture and Agricultural Science Procedia **8**: 243–251.

Brenner, N. and C. Schmid (2014). "The 'urban age' in question." International Journal of Urban and Regional Research **38**(3): 731–755.

Brinkerhoff, D. W., A. Wetterberg and E. Wibbels (2018). "Distance, services, and citizen perceptions of the state in rural Africa." Governance **31**(1): 103–124.

Cilliers, S., J. Cilliers, R. Lubbe and S. Siebert (2012). "Ecosystem services of urban green spaces in African countries—perspectives and challenges." Urban Ecosystems **16**(4): 681–702.

Cochrane, W. W. (2003). The curse of American agricultural abundance: a sustainable solution, U of Nebraska Press.

Cohen-Shacham, E., G. Walters, C. Janzen and S. Maginnis (2016). Nature-based solutions to address global societal challenges, Gland, Switzerland, IUCN.

Current, J., M. Daskin and D. Schilling (2002). Discrete network location models. Facility location: Applications and theory. Drezner Z. et al., 83–120. Springer.

Curry, N. and S. Owen (2009). "Rural planning in England: A critique of current policy." Town Planning Review **80**(6): 575–596.

Dandekar, H. C. (2015). "Rural planning: General." International Encyclopedia of the Social and Behavioral Sciences, January 1, 2002, 13425–13429.

Daskin, M. S. (1995). Network and discrete location: Models, algorithms and applications, John Wiley and Sons Inc., New York.

de Oliveira, E. S. and C. W. Thompson (2015). Green infrastructure and health. Handbook on green infrastructure: planning, design and implementation handbook on green infrastructure. D. Sinnett, N. Smith and S. Burgess, 11–29. Edward Elgar Publishing.

Dudzińska, M., S. Bacior and B. Prus (2018). "Considering the level of socio-economic development of rural areas in the context of infrastructural and traditional consolidations in Poland." Land Use Policy **79**: 759–773.

EEA (2015). Urban sustainability issues—What is a resource-efficient city? EEA Technical report Luxembourg: Publications Office of the European Union, European Environment Agency.

Ellegård, K. and U. Svedin (2012). "Torsten Hägerstrand's time-geography as the cradle of the activity approach in transport geography." Journal of Transport Geography **23**: 17–25.

Ely, M. and S. D. Pitman (2014). Green Infrastructure, Life support for human habitats: The compelling evidence for incorporating nature into urban environments South Australia, Department of Environment, Water and Natural Resources.

Filep-Kovács, K., Á. Sallay, Z. Mikházi, S. Jombach, Z. Szilvácsku, I. Valánszki and G. Gelencsér (2016). Green Infrastructure in Rural Development, Case Study in Hungary. Proceedings of the Fábos Conference on Landscape and Greenway Planning.

Gallent, N., M. Shucksmith and M. Tewdwr-Jones (2003). Housing in the European countryside: Rural pressure and policy in western Europe, Routledge.

Geurs, K. T., T. P. Dentinho and R. Patuelli (2016). Accessibility, equity and efficiency. eds., Edward Elgar, Accessibility, equity and efficiency, Edward Elgar Publishing: 3–8.

Geurs, K. T. and B. van Wee (2004). "Accessibility evaluation of land-use and transport strategies: review and research directions." Journal of Transport Geography **12**(2): 127–140.

Gkartzios, M., N. Gallent and M. Scott (2022). "A capitals framework for rural areas: 'Place-planning' the global countryside." Habitat International **127**: 1–10.

Gkartzios, M. and K. Scott (2015). "A cultural panic in the province? Counterurban mobilities, creativity, and crisis in Greece." Population, Space and Place **21**(8): 843–855.

Gleeson, B. (2013). "What role for social science in the 'Urban Age'?" International Journal of Urban and Regional Research **37**(5): 1839–1851.

Gray, M. L., C. R. Johnson and B. J. Gilley (2016). Queering the countryside: New frontiers in rural queer studies, NYU Press.

Harrison, P., K. Bobbins, C. Culwick, T.-L. Humby, C. L. Mantia, A. Todes and D. Weakley (2014). Urban resilience thinking for municipalities. Report Design and Layout: HotHouse South Africa, published by University of the Witwatersrand, Gauteng City-Region Observatory.

Hussain, S., R. Maqbool, A. Hussain and S. Ashfaq (2022). "Assessing the socio-economic impacts of rural infrastructure projects on community development." Buildings **12**(7): 1–18.

Hussain, S., W. Xuetong, R. Maqbool, M. Hussain and M. Shahnawaz (2022). "The influence of government support, organizational innovativeness and community participation in renewable energy project success: A case of Pakistan." Energy **239**(Part C). https://doi.org/10.1016/j.energy.2021.122172

Hussain, S., F. Zhu, Z. Ali and X. Xu (2017). "Rural residents' perception of construction project delays in Pakistan." Sustainability **9**(11). https://doi.org/10.3390/su9112108

Joubert, M. (2018). Identifying the potential of green infrastructure planning in rural and peri-urban informal settlements of South Africa, MSc., North-West University.

Kaufmann, V., M. M. Bergman and D. Joye (2004). "Motility: mobility as capital." International Journal of Urban and Regional Research **28**(4): 745–756.

Lapping, M. B. (2006). Rural policy and planning. Edited by Paul Cloke, Terry Marsden, Patrick Mooney, The handbook of rural studies. London, SAGE Publications Ltd.

Lapping, M. B. and M. Scott (2019). The evolution of rural planning in the Global North. The Routledge companion to rural planning, Routledge: 28–45.

Levine, J. (2020). "A century of evolution of the accessibility concept." Transportation Research Part D: Transport and Environment **83**. https://doi.org/10.1016/j.trd.2020.102309

Li, J. (2014). "Land sale venue and economic growth path: Evidence from China's urban land market." Habitat International **41**: 307–313.

Mamonova, N. and J. Franquesa (2019). "Populism, neoliberalism and agrarian movements in Europe. Understanding rural support for right-wing politics and looking for progressive solutions." Sociologia Ruralis **60**(4): 710–731.

Maqbool, R., Y. Rashid and S. Ashfaq (2022). "Renewable energy project success: Internal versus external stakeholders' satisfaction and influences of power-interest matrix." Sustainable Development **30**(6): 1542–1561.

Marcouiller, D. W., J. G. Clendenning and R. Kedzior (2002). "Natural amenity-led development and rural planning." Journal of Planning Literature **16**(4): 515–542.

Marsden, T. K., J. Murdoch, P. Lowe and N. Ward (2012). The differentiated countryside, Routledge.

Marshall, R. (2001). "Rural policy in the new century." International Regional Science Review **24**(1): 59–83.

Molla, M. B. (2015). "The value of urban green infrastructure and its environmental response in urban ecosystem: A literature review." International Journal of Environmental Sciences **4**(2): 89–101.

Monteiro, R., J. Ferreira and P. Antunes (2020). "Green infrastructure planning principles: an integrated literature review." Land **9**(12). https://doi.org/10.3390/land9120525

Mosher, A. T. (1976). Thinking about rural development, New York, Agricultural Development Council.

Natarajan, L. (2019). Planning for rural communities and major renewable energy infrastructure. The Routledge Companion to Rural Planning, Routledge: 548–556.

National Research Council (1989). Alternative agriculture, National Academies Press.

Naumann, S., M. Davis, T. Kaphengst, M. Pieterse and M. Rayment (2011). Design, implementation and cost elements of Green Infrastructure projects. Final report to the European Commission, Ecologic Institute and GHK Consulting.

Nelson, L. and P. B. Nelson (2011). "The global rural: Gentrification and linked migration in the rural USA." Progress in HumanGeography **35**(4): 441–459.

Osumgborogwu, I. (2016). "Analysis of access to social infrastructure in rural Imo State, Nigeria." Journal of Geography, Environment and Earth Science International **7**(2): 1–7.

Pichón, F. J., J. E. Uquillas and J. Frechione (1999). Traditional and modern natural resource management in Latin America, University of Pittsburgh Press.

Potter, R., T. Binns, J. Elliott, E. Nel and D. Smith (2018). Geographies of development: An introduction to development studies (4th ed.), Routledge.

Quintero, J. D. (2012). Principles, practices, and challenges for green infrastructure projects in Latin America, Washington, DC, Inter-American Development Bank.

ReVelle, C. S. and H. A. Eiselt (2005). "Location analysis: A synthesis and survey." European Journal of Operational Research **165**(1): 1–19.

Roe, M. and I. Mell (2013). "Negotiating value and priorities: evaluating the demands of green infrastructure development." Journal of Environmental Planning and Management **56**(5): 650–673.

Scott, M. J., N. Gallent and M. Gkartzios (2019). The Routledge companion to rural planning, Routledge London and New York.

Sharp, J. S., K. Agnitsch, V. Ryan and J. Flora (2002). "Social infrastructure and community economic development strategies: the case of self-development and industrial recruitment in rural Iowa." Journal of Rural Studies **18**(4): 405–417.

Shucksmith, M. and K. Rønningen (2011). "The Uplands after neoliberalism?—The role of the small farm in rural sustainability." Journal of Rural Studies **27**(3): 275–287.

Silva, J. A. (2019). "Green infrastructure in rural communities of Mexico." Cuadernos de Desarrollo Rural **16**(84): 1–16. doi: 10.11144/Javeriana.cdr16-84.girc

Sinakou, E., J. Boeve-de Pauw, M. Goossens and P. Van Petegem (2018). "Academics in the field of Education for Sustainable Development: Their conceptions of sustainable development." Journal of Cleaner Production **184**: 321–332.

Spatari, S., Z. Yu and F. A. Montalto (2011). "Life cycle implications of urban green infrastructure." Environ Pollut **159**(8–9): 2174–2179.

Szczygiel, B. and R. Hewitt (2000). "Nineteenth-century medical landscapes: John H. Rauch, Frederick Law Olmsted, and the search for salubrity." Bulletin of the History of Medicine **74**(4): 708–734.

Tan, Y., C. Shuai, L. Shen, L. Hou and G. Zhang (2019). "A study of sustainable practices in the sustainability leadership of international contractors." Sustainable Development **28**(4): 697–710.

Tonts, M. (2020). "Developmentalism, path dependence and multifunctionality: reflections on Australian rural planning cultures." Planning Theory & Practice **21**(5): 776–782.

USDA (2004). USDA Rural Development: committed to the future of America's rural communities. Rural cooperatives, United States, US Department of Agriculture. May/June 2004.

Villa, A. (2020). "Green infrastructure in Mexico: a booster for healthier cities." Deutsche Gesellschaft für Internationale Zusammenarbeit (GIZ). URBANET. Online https://www.urbanet.info/mexico-green-infrastructure/. Checked on, 2022.

Vitale Brovarone, E. (2021). "Accessibility and mobility in peripheral areas: a national place-based policy." European Planning Studies 30(8): 1444–1463.

Vitale Brovarone, E. and G. Cotella (2020). "Improving rural accessibility: a multilayer approach." Sustainability 12(7): https://doi.org/10.3390/su12072876

Wahid, A., M. S. Ahmad, N. B. Abu Talib, I. A. Shah, M. Tahir, F. A. Jan and M. Q. Saleem (2017). "Barriers to empowerment: assessment of community-led local development organizations in Pakistan." Renewable and Sustainable Energy Reviews 74: 1361–1370.

Ward Thompson, C. (2011). "Linking landscape and health: the recurring theme." Landscape and Urban Planning 99(3–4): 187–195.

3 Infrastructure and economy in rural and regional context

3.1 Introduction

Infrastructure systems are the backbones of our modern socioeconomic system. A well-connected infrastructures systems support the movement of goods and services for supporting multiple layers of users ranging from individuals to large corporations. A great level of accessibility to the infrastructure network is accomplished only when a continual and significant geographical movement of goods and services is maintained by effective management of hindrances and underlying externalities. Externalities may include possible unavailability of the labour forces, interrupted production from manufacturing plants, inconsistent economic growth, and market forces which may affect the underlying intents of positive value addition and overall contribution to community well-being (Carlsson, Otto et al. 2013). Due to the long lead time from planning and development to the operational phase of infrastructure projects, the functional performance and underlying support extended towards community well-being are not quite noticeable in the short term (Baldwin and Dixon 2008). Due to the interdependencies, the contribution to the economy from one type of infrastructure may be constrained by existing infrastructure systems (Batten and Karlsson 2012). The economic benefits of the infrastructure system especially in rural settings depend on numerous intrinsic factors such as socioeconomic, demographic, and even challenges and impacts resulting from climate change. Appropriate adjustments in infrastructures are necessary to enable, adapt, and enforce these factors. This argument highlights the need for careful considerations of necessary structural transformation and adaption of relevant economic growth models while planning and developing infrastructure for rural and regional requirements.

Growth theory is strongly intertwined with the theoretical study of the influence of infrastructure investment on economic development. For example, Arrow and Kurz (2013) incorporate infrastructure into the growth theory literature. As quantified by public capital, infrastructure has long been seen as a significant input in the production function. Barro (1990) studied the influence of public capital in the context of endogenous growth theory. Examining the influence of infrastructure development on economic growth

DOI: 10.1201/9781032622323-3

in Malaysia, Azam and Abu Bakar (2017) highlighted various predictors, including inbound foreign direct investment and change in human capital. The findings asserted that Malaysia's economic development in the 1990s was strongly influenced positively by the infrastructure boom. Inbound foreign direct investment and human capital are two key factors that are favourably correlated with economic growth. The study concluded infrastructure improvements are a core driver for encouraging economic expansion and increasing societal well-being. Considerable earlier empirical research including the studies by Futagami et al. (1993), Aschauer (1989), Easterly and Rebelo (1993), World Bank (1994), Sojoodi et al. (2012), and Palei (2015) demonstrates the critical impact of infrastructure in fostering economic growth and development.

Aside from the significance of infrastructures in supporting and executing fundamental transformation and growth, there has long been dispute regarding whether infrastructure expenditures truly cause economic growth. This theory has lately been accepted with the assumption that the issues of decarbonising the economy and encouraging economic growth may be addressed synergistically, with the rationale that infrastructure expenditures also generate economic growth (Carlsson, Otto et al. 2013). The economy of infrastructure is a long-debated topic, while many researchers argue that infrastructure investments certainly have a positive rate of return (Aschauer 1989, Macdonald 2008), some interpret the return of rate being the proportion of money earned or lost commensurate with the size of investments. Some researchers are more conservative, underestimating the infrastructures' significance in economic growth, and claiming that the accompanying consequences vary substantially depending on the type of infrastructure decisions made (Banister and Berechman 2003) or the quantity of supply already provided (International Monetary Fund 2005). Many studies are based on empirical data, with statistics used to estimate causation effects. No significant macroeconomic growth theory deals substantially with the economic implications of infrastructures (Carlsson, Otto et al. 2013). Infrastructure is frequently just one component of a more significant capital asset. Because different forms of capital serve different roles and have varied productivities, aggregation is problematic. As a result, infrastructure capital is sometimes classified as a distinct sort of capital meeting purpose-specific and need-based investment requirements. However, it is too common to presume that all infrastructure supply encourages growth. According to experts on the subject, the success of infrastructure expenditure in practice is, "at best mixed" (Henckel and McKibbin 2010) and "one size can't fit" for all the requirements. One of the fundamental distinctions between infrastructure capital and other types of capital is that it is susceptible to market imperfections and geographical impacts. Thus, the relative location of new infrastructure and connection in the network is essential (Braess 1968). This further dimension must be considered if new infrastructure is introduced or deleted from the network, whether purposefully or accidentally. Furthermore, infrastructure benefits from economies of scale (Holtz-Eakin and Lovely 1996), which means that cost advantages accrue as fixed expenses are dispersed over a greater output.

3.2   Economy-based infrastructure projects in rural development

There are numerous economy-based theories rationalising investments in infrastructure projects. While investment size and underlying return on investments are usually at the core of economy-based decisions, some of the macroeconomic growth theories focus on people's well-being and societal impacts while rationalising infrastructure projects with a community view-point (Glomm and Ravikumar 1994). While infrastructure services benefit human life including the potential for altering life expectancy in the long term, many of the existing economy-based theories are used for making decisions on infrastructure investments with a consideration of short-term profitability (Carlsson, Otto et al. 2013).

Over past decades however, there seems to be a shift in infrastructure investments due to increasing demand for understanding of impacts and limitations of publicly available infrastructure systems and the underlying influences on growth trajectories in the community (Fisher and Turnovsky 1998). While analysing demand, one of the broad concepts, *Congestion,* is often used to evaluate the cost of scarcity of infrastructure and costs involved in normalising the supply of goods and services in the community being considered. This is an essential part of economic thinking concerning infrastructure that is sometimes overlooked until scarcity in the infrastructure causes shortages, which eventually raises costs and, in extreme situations, prevents output entirely. Energy and water-related infrastructures make good examples in this context. The challenge with these ideas is often the ignorance of the significance of the spatial distributions and staggering requirements in service delivery (Carlsson, Otto et al. 2013). In classic macroeconomic growth theories, space is homogeneous, which means that a firm's location is unimportant because places in space do not have any distinguishing qualities such as differing local endowments (Carlsson, Otto et al. 2013).

The difficulty is not that the theories do not handle infrastructure capital, but that the economic models that underpin the models are intrinsically non-spatial. This is handled in modern neo-classical and endogenous growth models by incorporating geographical heterogeneities that oppose this essential feature (Behrens and Thisse 2007). Economic geography is the only type of macroeconomic theories, to some extent, considers spatiality (Krugman 1991, Fujita, Krugman et al. 2001), which shapes a chain of connected areas; yet, long-run growth mechanisms are not included in these models. However, this issue is later addressed by Cerina and Pigliaru (2007).

Anecdotally, none of the existing frameworks allows for an in-depth examination of the economic consequences of comprehensive infrastructure systems and/or infrastructure investments. While traditional cost-benefit analysis is used to rationalise large-scale expenditures, especially in public projects (Lakshmanan 2011), this method also falls short of the ultimate aim of understanding why an infrastructure asset is economically useful in the first place (Munnell 1992). Nonetheless, up to a certain extent, the cost-benefit analysis highlights the current economic practice's incapacity to account for agglomeration effects, reinforcing the necessity for macroeconomic growth theories that include spatial dimensions (Carlsson, Otto et al. 2013). In this regard, Carlsson et al. (2013) have considered this lacuna besides investigating the mechanisms through which infrastructures affect the rate of growth in the economy in a broader context. For instance, they demonstrated the critical importance of transportation and digital communications infrastructure in lowering trade costs and allowing economies of scale and knowledge accumulation (Carlsson, Otto et al. 2013). They also explored how poor infrastructure may constrain economic progress (Carlsson, Otto et al. 2013). Due to their intrinsic non-spatial character, they observed that some economic functions of infrastructure can be captured in existing macroeconomic models (Carlsson, Otto et al. 2013) (Image 3.1).

Referring to the broader benefits of the infrastructure systems for rural economy and community growth as depicted in Figure 3.1, much research-based evidence is available in the published literature. For instance, energy distribution to families and businesses allows for energy-dependent technologies in action, freeing time for people to engage in other activities (Buhr 2003). Energy infrastructure enables agents to consume energy created far from the place of production, allowing economic activity to cluster (Combes, Duranton et al. 2005, Carlsson, Otto et al. 2013).

Referring to the transportation infrastructure, it enables the transfer of population, supplies, services, and products across space, cutting shipping costs and alleviating the limits of local factor endowments, leading to a more dependency on economic growth to spatial structures rather than natural resources (Barbier 1999, Rodrigue 2020). Consequently, it permits drawing

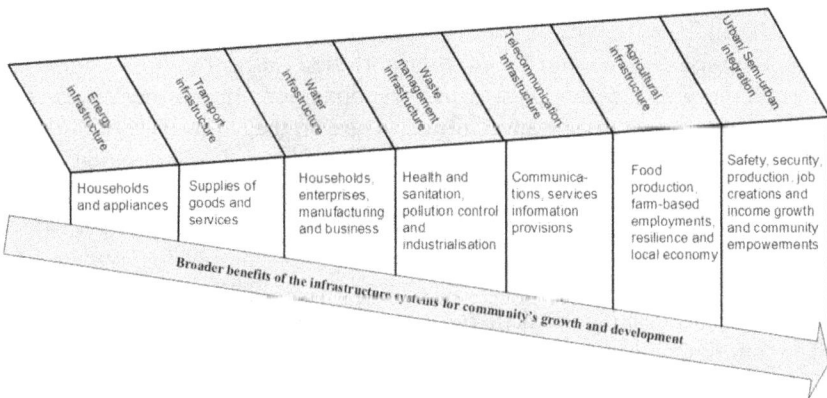

Figure 3.1 Infrastructure projects, rural economy, and growth trajectory.

Image 3.1 A tea garden supporting significant village economies in Assam.

more agents together to create more significant clusters (Fujita, Krugman et al. 2001). Transportation infrastructure also allows agents to access broader marketplaces for purchasing and selling products or recruiting taskforces, boosting competition, and converging prices. There have been many studies highlighting the role of transportation in economic growth and support for the general well-being of the community at large. For instance, Beyzatlar and Kustepeli (2011) employed yearly time series data covering the years 1950–2004 to examine historical relationships between railway infrastructure and the density of population as well as between railway infrastructure and economic growth in Turkey. The findings of the causality and co-integration tests indicated a long-term positive correlation between railroad length and real per capita GDP, as well as between railroad length and population density. While the length of the railway increased population density over the long and short terms, it only increased real per capita GDP over the long term. These empirical findings supported the theoretical assumption that, during the research period, improvements in transportation infrastructure resulted in higher incomes and better growth in the population. In another study, the effects of accessible transportation infrastructure on the regional economy have been assessed by Banerjee et al. (2020). They considered data from a period of 20 years and reported that in contrast to per capita GDP growth, their findings indicated that access to transportation networks has little impact on per capita GDP levels across industries (Banerjee, Duflo et al. 2020). Besides, to comprehend their findings, they developed a straightforward theoretical framework with testable predictions (Banerjee, Duflo et al. 2020).

The contribution of water infrastructure to economic growth among the agents is also widespread. While households may use the supply of water for household activities, enterprises may utilise it for business and manufacturing operations (Gatto and Lanzafame 2005). A reliable water supply enables people and businesses to function in regions without relying on natural and weather-dependent sources. This implies that a reliable supply of water enables businesses and homes to coexist. However, increasing the cost of water supply and usage restrictions may force the firms and people to be less dependent including encouraging them to migrate (Carlsson, Otto et al. 2013). Furthermore, water scarcity can interrupt manufacturing processes, lowering productivity and efficiency.

Wastewater management and solid waste disposal infrastructures are also vital to maintaining the sanitary condition of urban or rural areas. Corporations, households, and social venues rely on appropriate infrastructure for managing the waste and pollution they generate. If the supply of services is limited, negative externalities will accrue locally, causing health concerns and significant expenses while restricting the number of products being generated (Carlsson, Otto et al. 2013). If effective resource extraction plans are implemented, wastewater and solid waste may be used to extract resources, increasing total resource availability and overall resource utilisation efficiency in the economy (Carlsson, Otto et al. 2013). The degree of supply of water, waste, and sewage services is proportional to the health benefits of the community and thus appropriate supply commensurate with the needs of the community is important for achieving the expected benefits in health and wellbeing.

Telecommunications and internet infrastructures are the other important types of infrastructure providing fast communications, service consumption, and information provision for the agents. Therefore, they can facilitate access to a broader range of markets. With the recent trend towards working from home style, these infrastructures can also substitute commuting. They diminish the proclivity to consume physical items, which promotes energy efficiency and lowers the economy's resource intensity. A simple example is downloading software rather than purchasing a physical disk in traditional practices. Capacity restrictions may cause agents to lose access to data and marketplaces, resulting in inefficiencies and utility losses. Consumer technologies have been a growth mechanism in modern industrialised economies, co-evolving with the availability of telecommunication and internet infrastructures. To validate the importance of telecommunications advances, many researchers have conducted extensive studies around the world. For instance, Roller and Waverman (2001) evaluated how telecommunications infrastructure affects economic production for a panel of OECD nations over 20 years. The micro model for telecommunications investment and the macro production function revealed a considerable positive causal link, mainly when a critical mass of telecommunications infrastructure exists. Similarly, Esfahani and Ramrez (2003) discovered that the contribution of infrastructure services to GDP is extensive and, in general, exceeded the cost of providing those services for 75 countries during the decades 1965–1975, 1975–1985, and 1985–1995.

Similar to urban infrastructure, rural infrastructure raises agricultural production, boosts agricultural and non-farm employment, and raises the standard of life for rural residents, all of which contribute to rural economic growth and poverty reduction (Ghosh 2017). Although they are crucial for promoting agricultural investment and growth, roads, electrical supply, telecommunications, and other infrastructure services are claimed to be sparse in many rural regions (FAO 1996). It is also asserted that infrastructure services, such as roads for access to basic commodities for human life and healthy drinking water and sanitary to avoid disease, are crucial to people's welfare in terms of education and health (Datt and Ravallion 1998). Infrastructure deficiency is seen as a significant impediment to sustained human development (Ghosh 2017).

Increased productivity, improved access to marketplaces, higher consumer demand in rural areas, stimulation of the rural farm and non-agrarian economy, accelerated commercialisation of agriculture and the rural sector, and easier integration of rural economies with the national economy are all ways that rural infrastructure supports agricultural growth. High-yielding variety (HYV) technology adoption and diffusion heavily rely on rural amenities like irrigation, value-added services, loan accessibility, and marketing. The pivotal part that rural infrastructure plays in creating significant economic multiplier effects with an increase in agriculture has been reported by Mellor (1976).

Several studies such as Ellis and Nyasulu (2003), Jayaraman and Lanjouw (1999), Lanjouw et al. (2001), Reardon et al. (2001), and Zimmerman and Carter (2003) examined how increased infrastructure investment raises agricultural productivity and raises the standard of living for rural households by diversifying rural economic activities. According to Bhatia (1999), there is a significant positive correlation between India's food grain productivity per hectare and the rural infrastructure index, which includes rural electrification, roadways, transport, well-being, irrigation, farm credit, fertiliser, agrarian marketing, and research and extension.

Many researchers have also scrutinised how particular infrastructures, such as rural roads, transportation, energy, irrigation, and access to energy, influence agricultural development, and poverty reduction. For instance, Binswanger and his co-workers (1993) attempted to quantify the interactions between government, financial institutions, and farmer investment decisions, as well as their combined influence on agricultural investment and production. They stated that the rural banks and the availability of educational infrastructure significantly impact deciding on investments, inputs, and outputs (Binswanger, Khandker et al. 1993). According to their research, banks put their branches where the agro-climate and the infrastructure are advantageous for their operation, while farmers respond to infrastructure. Governments deploy their infrastructure investments in response to the agroclimatic potential of the districts. So, the mechanism by which farmers, the government, and intermediaries react to the same variables to influence agricultural production is intricate and dynamic (Binswanger, Khandker et al. 1993). Another study conducted by Chakraborty and Guha (2009) examined the relationship between economic growth and village-level connectivity in India. Rural roads were cited as the most significant infrastructure promoting rural development in the majority of surveys. By facilitating access to input and product markets and assisting farmers in realising higher input and output prices, rural roads assist in disseminating agricultural technology.

Rural roads were cited as the most significant infrastructure promoting rural development in the majority of research. By facilitating access to input and product markets and assisting farmers in realising higher input and output prices, rural roads assist in disseminating agricultural technology. By lowering the transaction and marginal costs of agricultural output, better road conditions result in more effective resource allocation (Ghosh 2017). Through trickle-down effects, increased agricultural productivity and production lessen rural poverty (Ghosh 2017). Through enhanced road and transportation infrastructure, rural households have easier access to financing, health care, and educational resources (Ghosh 2017). Better road connectivity always boosts backward and forward connections in the agricultural sector and enhances rural-urban ties (Ghosh 2017). This improves the poor people's quality of life by providing opportunities for jobs outside the hamlet (Ghosh 2017). Evidence from many nations' empirical studies also supports the significance of infrastructure for rural development (Fan and Zhang 2004, Li and Liu 2009, Estache, Wodon et al. 2014, Yamauchi 2016).

3.3 Significance of the need-based infrastructure and rural economy

Infrastructures could be defined as footings on which the building of the economy could be established. The formation of the gross development income (GDP) is a function of infrastructures (Sabir and Shamshir 2020). Besides, the so-called total factor productivity (TFP), which could be defined simply as the relationship between the involved inputs and the outputs in the production process, could be increased due to the decrease of trade costs as well as the increase of proficiency in terms of using other factor inputs of production (Sabir and Shamshir 2020). Investigating development and economic growth in China, Sahoo and his co-workers (2010) reported that the investment in the development of the infrastructures could help the creation of jobs for the lower segments of the society as well as enhancement of the economic activities. Besides, not only could the so-called physical (economic) infrastructures facilitate the access of the producers to the market and services effectively (Sabir and Shamshir 2020), but also it could lead to an ease of reciprocal access between the industry and the labour market as a result of which the unemployment rate could be dwindled. Apart from the mentioned advantages, an effective communication infrastructure could be a tool for the dissemination of information and ideas as well as the formation of startups, the former of which could help the investors and buyers to make informed decisions while the latter of which could provide new job opportunities. In addition to the physical infrastructures, social infrastructures could directly affect long-run economic growth as well as development (Easterly and Rebelo 1993, Fölster and Henrekson 2001). For instance, the investment in upstream elementary education is crucial for training the downstream workforce to support the rural economy. This investment in higher education (tertiary) could be an asset for the removal of skilled shortage as well as the production of more specialised professionals such as health professionals which could result in the improvement of health in the society at large.

For the study of the implication of infrastructure on the GDP, two different models i.e. neoclassical growth model and endogenous growth model could be used (Sabir and Shamshir 2020). These theories are different in context as the former states that sustained growth could only happen where technological advances are happening (technology-driven) and emphasises that the infrastructure investments could only have transient effects on the growth. However, the latter theory is based on the assumption that development occurs due to internal forces rather than external forces. In other words, it assumes that constant returns to scale or increased returns to scale correspond to capital and the non-transient growth could happen even without reliance on technology (Romer 1990, Rebelo 1991) or demographic factors (Becker and Barro 1988). After the study of Aschauer (1989) on the productivity of public expenditure for the United States' infrastructures, in which the fall of growth rate has been attributed to the fall of public expenditure on infrastructures, more attention has been drawn to the hypothesis that the infrastructures, productivity, and the economy are interrelated (Sabir and Shamshir 2020). However, it is worth mentioning that there are still major debates on how powerful the effect of public capital would be and what would be a reasonable assumption to draw logical conclusions on infrastructure spending and growth when it comes to rural communities. Figure 3.2 depicts the need-based infrastructure and the interconnected areas concerning the growth of the rural economy.

Figure 3.2 Need-based infrastructure and rural economy.

3.3.1 Interaction between physical and social infrastructures

There is a consensus that, given the right circumstances, investments in physical infrastructure help to reduce poverty and inequality through the channels of economic expansion and social development (Calderón and Servén 2008). Furthermore, physical and social infrastructure have been demonstrated to have various effects on social and economic development. For instance, improving the infrastructure for electricity has a significant influence on a company's productivity by lowering the production loss brought on by power surges and outages (Gnade, Blaauw et al. 2017). In addition, the protection and improvement of employee health led to an increase in productivity, which has a minor but still significant influence on a company's productivity (Gnade, Blaauw et al. 2017). Besides, numerous studies have discovered that physical infrastructure has a significant influence on how effectively social institutions such as educational and medical facilities operate (Brenneman and Kerf 2002). This is significant because the poor are negatively impacted more severely and persistently by urban-rural differences in access to healthcare and educational services (le R Booysen 2003). Investment in social and physical infrastructure is likewise linked to lessened economic inequality (Calderón and Servén 2008).

3.3.2 Linkages of Infrastructure spending with growth in the economy

Within the literature, clear linkages are evidenced between the investment in infrastructures and growth in the economy. In a research-based investigation, Aschauer (1989) and Barro (1990) hypothesised that economic development is eminent when infrastructure contributes to growth in marginal productivity in both private and public investment regimes (Sabir and Shamshir 2020). Agenor and Moreno-Dodson (2006) reviewed the various channels through which public infrastructures could affect economic growth and outlined that in addition to productivity economics, complementarity economics, and the so-called crowd-out effect theory are the concepts by which the effect of investment in infrastructures on the economy could be assessed. The former theory, e.g. complementarity theory has been introduced into the economics by Hicks (1932), stating that the demand for some goods will generate new demands for some associated goods. For example, a demand for a tabletop clock could lead to increased demands for batteries, or a demand for a printer could generate demand for toners. The latter theory, i.e. crowding out theory, states that the increase in public spending could lead to a decrease or even elimination of the private sector. Agenor and Moreno-Dodson (2006) have also mentioned the complementarity between private and public capital as well as the association between the reduced cost and capital reallocated between different sectors (Sabir and Shamshir 2020) both of which could impact the growth through investment adjustment cost. It is worth mentioning that to model the association between economic growth and infrastructures, researchers normally use various estimation techniques and adopt these on cross-sectional data, panel data, or time series data (Sabir and Shamshir 2020), which is beyond the scope of this book.

3.3.3 *Linkages of physical or social infrastructure spending with growth in disposable income*

Studies about the effect of infrastructures on disposable income are not quite widespread. Kumari and Sharma (2017) studied the relationship between the status of the physical and social infrastructure and the development in terms of the economy in the Indian context. Using the secondary data from the World Bank database, their research asserted that growth in gross domestic product (GDP) is directly proportional to the expenditures on primary and secondary education and health, telecommunication, air and rail transport, and power and energy. In contrast, however, Estache (2004) revealed little evidence regarding the direct influence of infrastructure on household income. Komives et al. (2001) highlighted how the demand for infrastructure varies as income rises as opposed to how infrastructure affects income. Estache et al. (2002) placed more emphasis on savings than income growth, leading to greater disposable income levels without any significant connection to the infrastructure expenditure. In contrast to raising household income directly, Brenneman and Kerf (2002) point out how basic and social infrastructure enhances the disposable income of families. The influence of physical and social infrastructure investments on disposable income may need further research-based investigation especially to draw a clear consensus among the rural communities.

3.4 Poverty reduction through rural infrastructure

Rural infrastructures have a positive influence on agrarian production, real incomes, and jobs in the agricultural and non-agrarian sectors, which contributes to poverty reduction (World Bank, 1994). Lower levels of poverty in rural and regional areas have been linked to higher farm production, per capita farm revenue, and employment in the agrarian and non-agrarian sectors. In this regard, infrastructures can directly help reduce poverty as they supply and facilitate the delivery of essential services like power access, clean potable water, and hygienic facilities. In addition, there is evidence that investing in rural infrastructure is linked to reduced transportation costs, reduced transaction costs for financial services, better access to markets for farmers, and a significant rise in agricultural output (Binswanger, Khandker et al. 1993). Numerous studies, including those by Fan and Hazell (1999), Fan and Hazell (2000), Fan et al. (1998), Fan et al. (2000a), Fan et al. (2000b), and Fan et al. (2000c), have examined the role of rural infrastructure in agricultural development, rural development, and the reduction of poverty in India and China. In addition, numerous studies have documented the beneficial impacts of rural infrastructure on rural development and poverty reduction through increased market size, economies of scale, better factor market performance, and commercialisation in the agricultural and rural sectors (Binswanger, Khandker et al. 1993, Jacoby 2001, McCleery 2005, Howe, Richards et al. 2019).

3.5 Effect of rural development on the livelihoods

Rural development planning is intertwined with rural livelihood. Perspectives on livelihoods begin with how various people live in various areas. According to the literature, there are several definitions available, such as the way of acquiring an income (Chambers 1995) or a mixture of the resources required and the actions conducted to survive (Scoones 2009). There is no clear pattern of how an individual could earn a living. If we represent individual earning patterns in rural settings in a relational mapping tool, the network of activities and relationships becomes highly complex. While such a pattern may go against traditional one-streamed income earning patterns in urban settings, by the virtue of the rural economy being intertwined across numerous related income-generating avenues, concentrate on specific activities like farming, wage employment, farm labour, local business, may not be a norm in most situations. However, the reality shows that people create a complicated bricolage of activities. Indeed, results might vary, but a key consideration in livelihood analysis is how various tactics alter the routes or trajectories of a person's livelihood. The concepts such as "coping", "adaptation", "improvement", "diversification", and "transformation" should be highlighted in a dynamic, longitudinal analysis (Scoones 2009). Thus, individual-level analysis is intricately linked to livelihood plans and family structure which again could differ from village to block or even district levels (Scoones 2009). To address complicated issues in rural development, understanding diversification as a livelihood solution is highly crucial. Due to the interdisciplinary nature of the income generation streams, the livelihood patterns of the individual members within the community provide a solid foundation for planning and development of the interrelated infrastructure systems, leveraging social and natural sciences. Livelihood methods are a perfect starting point for participatory research approaches in infrastructure planning since they focus on understanding complicated local realities including needs and priorities from both local and outsiders' perspectives (Scoones 2009).

Following the vigorous promotion of sustainable livelihood strategies in the 1990s (Carney 1999, Scoones 2002), many development organisations began to promote livelihood strategies as essential to their programming and even organisational structures. Some of the theoreticians of the livelihood perspectives claim that these perspectives can work at both micro and macro levels, while this is just an ambition (Scoones 2009). One of its fundamental flaws is the inadequacy of livelihood methods to address more extensive, global processes and how they affect local livelihood issues (Scoones 2009). Thus, livelihood views have frequently failed to participate in discussions

Image 3.2 Rice farms being prepared by individual farmers in the Monsoon Season who are dependent on rainwater.

about globalisation, abandoning the field of macroeconomics, which is infamous for being uninformed about local-level intricacies (Scoones 2009).

Dealing with long-term change presents livelihood perspectives with another difficulty. To be considered sustainable, a livelihood must be steady, long-lasting, robust, and resilient in the face of internal and external challenges (Scoones 2009). However, the livelihood perspectives have failed to address the sustainability of the livelihood in the long-term, instead, the focus has frequently been on coping and short-term adaptation, drawing on a rich tradition of vulnerability research (Swift 1989) (Image 3.2).

3.6 Effect of rural infrastructures on developing non-agricultural incomes

In the majority of rural places across the world, farming is the primary source of income for 90% of the people (Mphande 2016). In most such settings, rural communities often rely on small-scale farming, fishery, rearing livestock, and non-farm jobs. In many situations, rural inhabitants frequently live in poverty and are deprived of the essential need of comfortable living which in turn presents a significant barrier to a rural existence. Many times, single income is the sole source of survival in rural families. Thus, compared to a rural home with only one source of income, one with a variety of sources has a higher chance of surviving financially. Therefore, a variety of training and education for skill and career development opportunities are fundamental ingredients for achieving reliable rural livelihood.

Literacy is a transferrable skill that can be leveraged for learning and earning over a long period. A person's chances of selecting an acceptable career path or pursuing more education to improve their employability are increased if they can read and write. The majority of knowledge for developing skills and applications that can provide livelihoods comes from written communication and verbal interactions. While this is true for most of the common public, the need for people with disability may even be greater in rural areas. With sufficient literature skills, access to credit and financing that may be utilised as startup capital or pursuing continuous career development through specific vocational pathways. In this regard, access to educational facilities which are parts of social infrastructure systems could improve the livelihood of the rural households.

The availability of financial institutions as well as proper infrastructures to provide access to low-rate loans and lines of credit are the other key points for improving the livelihoods of the rural community. Having access to credit is advantageous to a community because it provides the funds needed to launch a new enterprise or purchase assets that may be utilised to improve a living. Diversification, however, must be supported by the capability to secure collateral and repayment of the loan; otherwise, poverty levels have risen as a result of lost assets due to loan default.

People who live in rural areas adjacent to urban areas have access to infrastructures like marketplaces, banks, credit establishments, and medical facilities that may help them advance their income. They can also find markets for their products and underlying liquidation opportunities. A household's ability to access non-farm occupations and skills is also facilitated by proximity to a town and accessibility through interconnected infrastructure networks. More remote homes or communities have lesser access to these services and have a lower likelihood of diversifying.

Human capital is valuable in rural income diversification as it simplifies a household's livelihood. In other words, acquiring more skills by household members will lead to the expansion of their capacity to explore other marketplaces as well as the generation of a more sustainable income for the family. Besides, the diversification due to the increase of human capital and self-development could lead to a better social status and capital for the individuals.

Despite the advantages of the diversification of income for the rural community, constraints remain that could impede the process of diversification. For instance, lack or scarcity of proper infrastructures either physical or social could lead to the isolation of the rural community and deprive them of having the opportunity to develop themselves or access to proper or competitive markets. Lack of knowledge might cause households to pass up chances that could improve their professional growth and job prospects. Limited access or lack of access to financial institutions will lead to a reluctance amongst households to pursue diversification of their income as the risk of failing in a new market might seriously harm their financial status. Lack of awareness is the other impediment constraining diversification as it

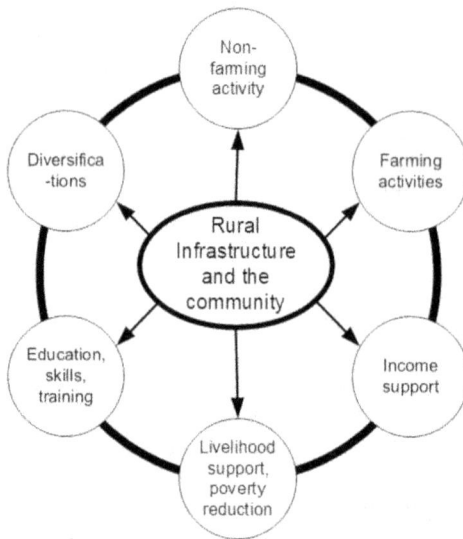

Figure 3.3 Rural infrastructure and support areas.

can bring about missing an opportunity that could be economically viable and advantageous. The degree of diversity can also be impacted by the absence of assets like cash, real estate, and residences that can be sold to generate additional income for a household or used as collateral for loans. Diversifying sources of income for some rural and regional districts might be cumbersome due to the climatic conditions. Investors may stay away from regions with severe weather. The absence of the required infrastructure prevents the growth of new industries and markets in these areas. Figure 3.3 highlights the interconnected network of support activities associated with rural infrastructure systems.

3.7   Infrastructure and rural economy

The above discussions highlight how the infrastructure system is closely linked to the development of the rural community by supporting the needs and requirements of amenities across numerous fronts. It is also apparent that due to be virtue of rural infrastructure being location-specific interventions for catering primarily for the needs of the local communities, the traditional rationalisation theories of infrastructure investments are not often applicable when it comes the decision-making in rural settings. What works in urban areas may not work in rural regions. As the "rural economy" in rural settings is entirely different from the "money economy" in urban settings, the decisions on infrastructure investments need to take into consideration of variety of different typologies related to the rural conditions, culture and heritage, opportunities and potentials offered by the geographical boundaries including demographic profile and density, etc.

Before embarking on rural infrastructure planning, an in-depth understanding of the rural economy is essential. While rural and urban relations are a topic of long debate among social scientists and policymakers, there is no clear consensus on what constitutes a rural region and the underlying characteristics influencing the rural economy. Many researchers tend to define rural as a mere byproduct of what can't be fitted within the definition of a city. In the context of infrastructure planning and investment where money plays a big role, the status of the countries whether capitalistic or socialist becomes the important determinant. For instance, with capitalist viewpoints, the rural regions are not quite considered significant as the space or place is viewed from business or industrialisation perspectives. In a capitalist society, rural or remote regions are seen as hindrances in the production of goods or services which are important measures in gross national income and also perhaps GDP growth. However, from a socialistic viewpoint, rural or regional areas are viewed as nature-based potentials with a particular emphasis on factors such as environmental conservation, and preservation of culture and heritage with distinctive needs and requirements concerning the purpose of the place. This book focuses on the infrastructure needs for smart villages with a particular emphasis on the rural or remote community, the underlying principle for driving rural infrastructure needs is considered to be entirely different from the city. The remainder of the book will expand the discussions on the premise of rural planning and decision-making focusing on the rural community (Image 3.3).

Image 3.3 An artistic observation tower in the wetlands in a village in Assam, for supporting the non-agricultural economy.

3.8　Summary

Concerning some of the unique literature on rural infrastructure and economy, this chapter discusses the key issues and considerations associated with the planning and development of infrastructure in rural settings. One of the key arguments this chapter presents is the significance of infrastructure in rural development and community empowerment. Contrasting to the traditional practices of decision-making in urban settings, the literature-based scientific discussions highlight the fact that rural infrastructure is not always driven by economic rationale. Rather rural infrastructure is the lifeline for addressing the community-specific challenges such as agricultural and non-agricultural support, provisions for livelihood and poverty eradication, facilitating education, skill and training, income generation and diversification of the opportunities for maximising growth potentials among the community.

The next chapter will look in detail at the decision-making of the rural infrastructure systems from a planning and investment perspective. In infrastructure decision-making, the whole-of-life approach is another important consideration encompassing the entire lifecycle of projects.

References

Agénor, P.-R. and B. Moreno-Dodson (2006). Public infrastructure and growth: New channels and policy implications, World Bank Publications, **4064**.

Arrow, K. J. and M. Kruz (2013). Public investment, the rate of return, and optimal fiscal policy, RFF Press.

Aschauer, D. A. (1989). "Is public expenditure productive?" Journal of Monetary Economics **23**(2): 177–200.

Azam, M. and N. A. A. Bakar (2017). "The role of infrastructure in national economic development: Evidence from Malaysia." Journal of Economic & Management Perspectives **11**(4): 630–637.

Baldwin, J. R. and J. Dixon (2008). "Infrastructure capital: What is it? Where is it? How much of it is there?" Canadian Productivity Review Research Paper No. 16.

Banerjee, A., E. Duflo and N. Qian (2020). "On the road: Access to transportation infrastructure and economic growth in China." Journal of Development Economics **145**(C). doi: 10.1016/j.jdeveco.2020.102442

Banister, D. and J. Berechman (2003). Transport investment and economic development, Routledge.

Barbier, E. B. (1999). "Endogenous growth and natural resource scarcity." Environmental and Resource Economics **14**(1): 51–74.

Barro, R. J. (1990). "Government spending in a simple model of endogeneous growth." Journal of Political Economy **98**(5, Part 2): S103–S125.

Batten, D. F. and C. Karlsson (2012). Infrastructure and the complexity of economic development, Springer Science & Business Media.

Becker, G. S. and R. J. Barro (1988). "A reformulation of the economic theory of fertility." The Quarterly Journal of Economics **103**(1): 1–25.

Behrens, K. and J.-F. Thisse (2007). "Regional economics: A new economic geography perspective." Regional Science and Urban Economics **37**(4): 457–465.

Beyzatlar, M. A. and Y. R. Kustepeli (2011). "Infrastructure, economic growth and population density in Turkey." International Journal of Economic Sciences and Applied Research **4**(3): 39–57.

Bhatia, M. S. (1999). "Rural infrastructure and growth in agriculture." Economic and Political Weekly **34**(13): A43–A48.

Binswanger, H. P., S. R. Khandker and M. R. Rosenzweig (1993). "How infrastructure and financial institutions affect agricultural output and investment in India." Journal of Development Economics **41**(2): 337–366.

Braess, D. (1968). "Über ein Paradoxon aus der Verkehrsplanung." Unternehmensforschung **12**(1): 258–268.

Brenneman, A. and M. Kerf (2002). "Infrastructure & poverty linkages." A Literature Review, Washington, DC, The World Bank.

Buhr, W. (2003). "What is infrastructure?" Volkswirtschaftliche Diskussionsbeiträge. Universität Siegen, Fakultät Wirtschaftswissenschaften, Wirtschaftsinformatik und Wirtschaftsrecht.

Calderón, C. and L. Servén (2008). Infrastructure and Economic Development in Sub-Saharan Africa, World Bank Policy Research Working Paper No. 4712.

Carlsson, R., A. Otto and J. W. Hall (2013). "The role of infrastructure in macroeconomic growth theories." Civil Engineering and Environmental Systems **30**(3–4): 263–273.

Carney, D. (1999). Livelihoods approaches compared: A brief comparison of the livelihoods approaches of the UK Department for International Development (DFID), CARE, Oxfam and the United Nations Development Programme (UNDP), Department for International Development.

Cerina, F. and F. Pigliaru (2007). Chapter 5: Agglomeration and growth in the NEG: A critical assessment. New directions in economic geography. B. Fingleton, Edward Elgar Publishing.

Chakraborty, D. and A. Guha (2009). "Infrastructure and economic growth in India." Journal of Infrastructure Development **1**(1): 67–86.

Chambers, R. (1995). "Poverty and livelihoods: whose reality counts?" Environment and Urbanization **7**(1): 173–204.

Combes, P. P., G. Duranton and H. G. Overman (2005). "Agglomeration and the adjustment of the spatial economy." Papers in Regional Science **84**(3): 311–349.

Datt, G. and M. Ravallion (1998). "Why have some Indian states done better than others at reducing rural poverty?" Economica **65**(257): 17–38.

Easterly, W. and S. Rebelo (1993). "Fiscal policy and economic growth." Journal of Monetary Economics **32**(3): 417–458.

Ellis, F., M. Kutengule and A. Nyasulu (2003). "Livelihoods and rural poverty reduction in Malawi." World Development **31**(9): 1495–1510.

Esfahani, H. S. and M. a. T. Ramírez (2003). "Institutions, infrastructure, and economic growth." Journal of Development Economics **70**(2): 443–477.

Estache, A. (2004). Emerging infrastructure policy issues in developing countries: a survey of the recent economic literature.

Estache, A., V. Foster and Q. Wodon (2002). Accounting for poverty in infrastructure reform.

Estache, A., Q. Wodon and K. Lomas (2014). Infrastructure and poverty in sub-Saharan Africa, Springer.

Fan, S. and P. Hazell (2000). "Should developing countries invest more in less-favoured areas? An empirical analysis of rural India." Economic and Political Weekly **35**(17): 1455–1464.

Fan, S., P. Hazell and T. Haque (2000a). "Targeting public investments by agro-ecological zone to achieve growth and poverty alleviation goals in rural India." Food Policy **25**(4): 411–428.

Fan, S., P. Hazell and S. Thorat (2000b). "Government spending, growth and poverty in rural India." American Journal of Agricultural Economics **82**(4): 1038–1051.

Fan, S., P. Hazell and S. K. Thorat (2000c). "Impact of public expenditure on poverty in rural India." Economic and Political Weekly **35**(40): 3581–3588.

Fan, S. and P. B. Hazell (1999). Are returns to public investment lower in less-favored rural areas?: an empirical analysis of India.

Fan, S., P. B. Hazell and T. Haque (1998). "Role of infrastructure in production growth and poverty reduction in Indian rainfed agriculture." Project report to the Indian Council for Agricultural Research and the World Bank, Washington, DC, USA, International Food Policy Research Institute.

Fan, S. and X. Zhang (2004). "Infrastructure and regional economic development in rural China." China Economic Review **15**(2): 203–214.

FAO (1996). "Technical background documents." World Food Summit. Rome, Food and Agriculture Organization of the United Nations **2**: 6–11.

Fisher, W. H. and S. J. Turnovsky (1998). "Public investment, congestion, and private capital accumulation." The Economic Journal **108**(447): 399–413.

Fölster, S. and M. Henrekson (2001). "Growth effects of government expenditure and taxation in rich countries." European Economic Review **45**(8): 1501–1520.

Fujita, M., P. R. Krugman and A. Venables (2001). The spatial economy: Cities, regions, and international trade, MIT press.

Futagami, K., Y. Morita and A. Shibata (1993). "Dynamic analysis of an endogenous growth model with public capital." The Scandinavian Journal of Economics **95**(4): 607–625.

Gatto, E. and M. Lanzafame (2005). "Water resource as a factor of production-water use and economic growth." In the proceedings of the 45th Congress of the European Regional Science Association (ERSA) Conference, Eds. Rietveld P, University of Amsterdam, The Netherlands, August 2005.

Ghosh, M. (2017). "Infrastructure and development in rural India." Margin: The Journal of Applied Economic Research **11**(3): 256–289.

Glomm, G. and B. Ravikumar (1994). "Public investment in infrastructure in a simple growth model." Journal of Economic Dynamics and Control **18**(6): 1173–1187.

Gnade, H., P. F. Blaauw and T. Greyling (2017). "The impact of basic and social infrastructure investment on South African economic growth and development." Development Southern Africa **34**(3): 347–364.

Henckel, T. and W. McKibbin (2010). "The economics of infrastructure in a globalized world: issues, lessons and future challenges." Washington DC, The Brookings Institution, **10**.

Hicks, J. R. (1932). "Marginal productivity and the principle of variation." Economica (35): 79–88.

Holtz-Eakin, D. and M. E. Lovely (1996). "Scale economies, returns to variety, and the productivity of public infrastructure." Regional Science and Urban Economics **26**(2): 105–123.

Howe, J., P. Richards and J. Howe (2019). Rural roads and poverty alleviation, Routledge.

International Monetary Fund (2005). Public investment and fiscal policy—Lessons from the pilot country studies, Fiscal Affairs Department.

Jacoby, H. G. (2001). "Access to markets and the benefits of rural roads." The Economic Journal **110**(465): 713–737.

Jayaraman, R. and P. Lanjouw (1999). "The evolution of poverty and inequality in Indian villages." The World Bank Research Observer **14**(1): 1–30.

Komives, K., D. Whittington and X. Wu (2001). Infrastructure coverage and the poor: A global perspective, World Bank Publications.

Krugman, P. (1991). "Increasing returns and economic geography." Journal of Political Economy **99**(3): 483–499.

Kumari, A. and A. Kumar Sharma (2017). "Infrastructure financing and development: A bibliometric review." International Journal of Critical Infrastructure Protection **16**: 49–65.

Lakshmanan, T. R. (2011). "The broader economic consequences of transport infrastructure investments." Journal of Transport Geography **19**(1): 1–12.

Lanjouw, P., J. Quizon and R. Sparrow (2001). "Non-agricultural earnings in peri-urban areas of Tanzania: evidence from household survey data." Food Policy **26**(4): 385–403.

le R Booysen, F. (2003). "Urban–rural inequalities in health care delivery in South Africa." Development Southern Africa **20**(5): 659–673.

Li, Z. and X. Liu (2009). The effects of rural infrastructure development on agricultural production technical efficiency: evidence from the data of Second National Agricultural Census of China. 2009 Conference, August 16-22, 2009, Beijing, China, International Association of Agricultural Economists. doi: 10.22004/ag.econ.51028

Macdonald, R. (2008). An Examination of Public Capital's Role in Production. Micro-economic Analysis Division, 18-F, R.H. Coats Building, 100 Tunney's Pasture Driveway Statistics Canada, Ottawa K1A 0T6.

McCleery, S. J. R. (2005). Making infrastructure work for the poor. New York, NY, United Nations Development Program.

Mellor, J. W. (1976). The new economics of growth; a strategy for India and the developing world. USA: Cornell University Press.

Mphande, F. A. (2016). Rural livelihood. Infectious diseases and rural livelihood in developing countries. Singapore, Springer Singapore: 17–34.

Munnell, A. H. (1992). "Policy watch: infrastructure investment and economic growth." Journal of Economic Perspectives **6**(4): 189–198.

Palei, T. (2015). "Assessing the impact of infrastructure on economic growth and global competitiveness." Procedia Economics and Finance **23**: 168–175.

Reardon, T., J. Berdegué and G. Escobar (2001). "Rural nonfarm employment and incomes in Latin America: overview and policy implications." World Development **29**(3): 395–409.

Rebelo, S. (1991). "Long-run policy analysis and long-run growth." Journal of Political Economy **99**(3): 500–521.

Rodrigue, J.-P. (2020). The geography of transport systems, Routledge.

Roller, L.-H. and L. Waverman (2001). "Telecommunications infrastructure and economic development: A simultaneous approach." American Economic Review **91**(4): 909–923.

Romer, P. M. (1990). "Human capital and growth: Theory and evidence." Carnegie-Rochester Conference Series on Public Policy **32**: 251–286.

Sabir, S. and M. Shamshir (2020). "Impact of economic and social infrastructure on the long-run economic growth of Pakistan." Sustainable Water Resources Management **6**(1). https://doi.org/10.1007/s40899-020-00361-3

Sahoo, P., R. K. Dash and G. Nataraj (2010). "Infrastructure development and economic growth in China." Institute of Developing Economies Discussion Paper **261**.

Scoones, I. (2002). Sustainable rural livelihoods: a framework for analysis. working paper. Brighton, England, Institute of Development Studies, **72**.

Scoones, I. (2009). "Livelihoods perspectives and rural development." The Journal of Peasant Studies **36**(1): 171–196.

Sojoodi, S., F. Mohseni Zonuzi and N. Mehin Aslani Nia (2012). "The role of infrastructure in promoting economic growth in Iran." Iranian Economic Review **16**(32): 111–132.

Swift, J. (1989). "Why are rural people vulnerable to famine?" IDS Bulletin **20**(2): 8–15.

World Bank (1994). World Development Report 1994: Infrastructure for Development. World Development Report 1994. New York, Oxford University Press.

Yamauchi, F. (2016). "The effects of improved roads on wages and employment: evidence from rural labour markets in Indonesia." The Journal of Development Studies 52(7): 1046–1061.

Zimmerman, F. J. and M. R. Carter (2003). "Asset smoothing, consumption smoothing and the reproduction of inequality under risk and subsistence constraints." Journal of Development Economics 71(2): 233–260.

4 Decision-making on infrastructure in rural and regional context

4.1 Introduction

The rationale for decision-making on infrastructure in rural and regional settings is vastly different from the urban centres. While decision-making in the urban infrastructure could be based on factors such as economy of scale, competitiveness, market accessibility, etc., decision-making in the rural regions may focus on place-based priorities in the context of challenges and opportunities at a global scale. Long physical distances of the rural regions from the city centres along with a lack of competitive market forces and specificity in the production of goods and services make the traditional rationalisation of investment decisions quite redundant.

As rural planning and development is closely coupled with the general welfare and livelihood of the rural community, the lowest denomination of the decision-making relies on factors such as demographic data including needs and requirements for supporting empowerment and improvisations of the community. Many development agencies including governments in developing countries consider rural infrastructure as a key enabler for meeting the basic needs of the community and closing the gap in regards to the thresholds set out in the UN's sustainable development goals. While the traditional practice for making decisions on rural infrastructure is mostly top-down where scarce funds are always an issue for rural planning and development, due to the rapid decentralisation of funds and also generally the governance structure in governments of many developing countries, rural infrastructure is certainly receiving much greater attention over past decades. As the governance gets decentralised, so as the actors and the policy-makers. Thus, shifts from the top-down bureaucratic processes to more bottom-up and community-centric approaches are increasingly being realised for making effective planning and development decisions in rural settings.

4.2 Challenges in rural investment decisions

Over the past few decades, structural change in agricultural and rural regions has increased due to general economic, political, and sociological factors,

DOI: 10.1201/9781032622323-4

having favourable and unfavourable consequences. Rural marginalisation refers to negative trends that are particularly noticeable in remote and underdeveloped areas (Wiesinger 2007). These trends include socio-economic and cultural decline, out-migration, brain drain, poverty in rural and regional districts and social exclusion, loss of infrastructure and services, environmental degradation, biodiversity loss, habitat loss, deforestation, land abandonment, and landscape degradation (Wiesinger 2007). Rural marginalisation has long been the focus of rural and regional development and planning, rural sociology research, and rural policy studies. These problematic, undesirable, and frequently unintended incidents are caused by local and international political factors (such as a lack of or inadequate rural development initiatives, globalisation, and telecommunication), as well as by so-called intrinsic aspects that correspond with local communities' social structures and sociological patterns (Wiesinger 2007). Rural marginalisation can be primarily attributed to unfavourable circumstances and a lack of resources. However, this rule does not hold in all regions or entirely. Some very remote areas with few policy initiatives, weak economies, and unfavourable climatic circumstances turn out to be more resilient than others that are substantially better off. The marginalisation dynamic appears to include some sort of intangible asset. This indicator econometrics conundrum necessitates a different strategy. Social capital has been proposed as a tool or missing link in several recent papers to explain this relatively sophisticated and intricate connection (Wiesinger 2007).

Public infrastructure and services are being neglected throughout this period of globalisation and neo-liberalism, which is also considered the dominant economic paradigm (Wiesinger 2007). This lack of interest negatively impacts isolated rural regions, the rural economy, and the entire social fabric. People who are socially vulnerable and disadvantaged suffer the most when public assistance is reduced. As a result, rural poverty and social isolation are spreading worldwide. Consequently, considering the intrinsic capabilities of the rural areas and underlying potentials, nature-based development started gaining some popularity in both political and economic contexts. Mutual aid, self-consciousness, and empowerment were viewed as a cure in thriving rural communities with active regional and local political engagement (Wiesinger 2007). In this context, locals are encouraged to rely more on their capabilities but without much dependence on public assistance. Following are some important social considerations that warrant infrastructure investment decisions in rural settings.

4.2.1   *Social protection and social graduation*

The social protection concept has been increasingly considered as one of the key basic requirements addressing the vulnerabilities and risks associated with economic activities, and social isolation among the rural communities. This concept has been employed at the core of public policies or programs supporting infrastructure investment and targeting poverty alleviation and family protection. Roelen and Devereux (2013) advocate that relevant infrastructure is the key to promoting an inclusive framework and achieving social protection. They argued that appropriate infrastructure provisions are needed to ensure universal access to social protection of the community at large and address the structural causes of poverty and vulnerability at the grassroots level (Roelen and Devereux 2013). Following the ideals of inclusive social and economic growth supported by proactive decisions on relevant and targeted infrastructure provisions, an inclusive approach to social protection would not target a lucky few for a certain period but the community at large (Roelen and Devereux 2013). It would support all sections of the community based on the needs and requirements in the localised settings.

In another study, Sabates-Wheeler and Devereux (2011) introduced a framework to facilitate the so-called graduation in social protection programs. "Graduation" in social protection programs is the term used to describe how social protection enables individuals to escape poverty and to do so indefinitely without the need for ongoing transfers (Roelen and Devereux 2013). They reported that this transformative framework could offer a more proactive and constructive role for social protection (Sabates-Wheeler and Devereux 2007). They further added that this framework also could allow for discovering significant synergies between the "economic" (provision, prevention, and promotion) and social (transformation) activities provided by various social protection programs (Sabates-Wheeler and Devereux 2007). A transformational approach broadens the notion of social protection beyond targeted income and expenditure transfers to include chronic poverty and risks to livelihoods (Sabates-Wheeler and Devereux 2007).

4.2.2 Social security and welfare

International labour organisations defined social protection as a broad concept that can cover social security as well as non-statuary schemes (International Labour Office 2001). Some researchers have been also using the term "social security" as a synonym for social protection for many years (Midgley 2012). However, scholars such as Standing (2007) rejected the notion that the terms "social protection", "social security", and "welfare" are synonyms. He states that the most comprehensive kind of protection is social protection, which refers to the whole spectrum of institutional protections, protective transfers, and services designed to safeguard the at-risk population (Standing 2007). However, social security and welfare are at least one level higher which require completely different considerations when it comes to facilitating infrastructure provisions and underlying support mechanisms. While the goals for social protection could be viewed as a basic human entitlement where the level of intervention levels may vary from one location to the next, social security and welfare provisions are set based on considerations such as demographic and socio-economic profiles of the specific community. Midgley (2012) asserted poverty elimination and income protection as the goals of social protection, while Devereux and Sabates-Wheeler (2004) consider the vulnerability reduction for poor people as the focus of social protection. Humanitarian principles, social inclusion, and the dignity of humans were also other stated objectives of social protection (International Labour Office 2001). The chronically poor, the economically vulnerable, and the socially outcast are three groups of persons who can benefit from social protection, according to Devereux and Sabates-Wheeler (2004). Additionally, the so-called social protection floor strategy that consists of "basic services and social transfers" benefits both emerging and developed nations (ILO and WHO 2009). According to the International Labour Office (2011), a social protection floor strategy is the first step in protecting the most vulnerable citizens in countries in the Global South. However, in the Global North, where the social security welfare system is well established, this system is necessary to address the coverage gap by reaching vulnerable populations (Sampson and Drolet 2016) (Image 4.1).

4.2.3   Social insurance and social assistance

While social insurance and social assistance are two main components of social protection (Sampson and Drolet 2016), these two dimensions demand a slightly different consideration of policy provision and implementation plans. The former is practically identical to social security, in which people or families employ the insurance concept to protect themselves against dangers by pooling their resources with those experiencing

Image 4.1 Concrete-lined Well for potable water solution within the social infrastructure system.

comparable vulnerabilities (Economic and Social Council of United Nations 2001). Social insurance is funded via monthly contributions made by persons participating in programs such as pensions, severance pay, health insurance, and unemployment compensation. However, these initiatives are frequently restricted to a small number of formal-sector employees. The latter includes government-sponsored and non-governmental public acts that aim to transfer resources to groups of persons deemed eligible owing to their deprived status (Conway, De Haan et al. 2000). Some researchers also suggested that social legislation, as well as social services, could empower the marginalised residents of rural areas (Gentilini and Omamo 2011). Besides, legislative and social laws could be enforced by the government or be followed by non-governmental organisations, private enterprises, or even civil society.

In terms of providing support and services to reduce vulnerabilities and eliminate hazards that, if not handled, might result in poverty or further entrench those who are already poor, policies and programs are seen as a crucial component of social protection (Sampson and Drolet 2016). Ortiz (2001) believes that plans and policies for the labour market that aid in stabilising and adjusting the labour market; social insurance schemes that act as a buffer against threats to one's way of life due to illness, unemployment, and accidents at work; public care and aid that provide the minimal requirements of those in need who have no other sources of support; protection against threats like crop losses or interruptions through agriculture insurance; and community-based social funds and employment initiatives that provide folks with short-term employment opportunities, are five key activities that are closely related to the social protection.

In general, the focus of social protection is centred on risks and vulnerabilities that can affect human development. Therefore, it includes a multifaceted strategy to protect people and their families from shocks, stresses, and deprivation through various interventions that improve livelihood security, social inclusion, and, most importantly, human dignity (Sampson and Drolet 2016). Implementation of the relevant strategies for addressing the above issues requires careful planning of services and infrastructure provisions concerning the particular community's needs and commensurate with the demographic profile and geographic location.

4.2.4 Backcasting and community participation

Backcasting is one of the methods that can be used to engage stakeholders in the process of establishing long-term strategies and is proven to engage more participants than the traditional participatory tools (Sisto, Fernández-Portillo et al. 2022). Backcasting incorporates people in every step of the strategic planning process. It has been applied to planning in uncertain conditions with positive outcomes that might raise the possibility of handling complex problems in a methodical and coordinated manner

(Robinson, Burch et al. 2011, Sisto, Fernández-Portillo et al. 2022). Furthermore, backcasting allows for the inclusion of local community and stakeholder information in the outcomes because of its participatory character (Sisto, Lopolito et al. 2018). Additionally, it provides a chronology that can aid policymakers in better planning their future course of action (Sisto, Lopolito et al. 2018). In this method, an ideal future is first defined, and then steps that will link that future to the present are outlined backwards (Robinson 2003, van Vliet and Kok 2013). Backcasting, in particular, focuses on every potential solution to address ongoing and incoming issues to achieve the desired future rather than attempting to anticipate the future. According to this perspective, this method is especially well suited to participatory strategic planning, where policy measures are chosen based on external factors, such as their attractiveness in terms of the social or environmental context (Robinson 2003).

The history of using backcasting shows that its first applications were for the energy sector (Lovins 1976). However, the focus then shifted to sustainability issues (Robinson, Burch et al. 2011) and rural strategic planning and decision-making (Sisto, Lopolito et al. 2018).

To outline a development plan for a rural and regional area under the LEADER to Community-Led Local Development (LEADER-CLLD) approach, it is crucial to recognise the actions through a participatory process that is both in line with the methodological recommendations of the European Commission and amenable to lowering the risks of policy failure (Sisto, Fernández-Portillo et al. 2022). Besides, it is also essential to choose the best alternative solutions ensuring delivery of goods and services as per the targets and objectives (Sisto, Fernández-Portillo et al. 2022).

4.3 Theories and practices behind regional policy and decision support

Rural regions are a complex network of historical, social, and political components in developed and developing nations (Terluin 2003). Promoting new business creation and new development pathways in these sectors necessitates a decision-making process typically hampered by various forms of uncertainty. Lack of understanding of human behaviour, inherent variability, the unpredictability of such behaviour, and diversity in interest and its results on the interactions between the actors are some of the sources of this uncertainty. This stochastic nature could make the final choice even more challenging and consequently lead to a wrong or more difficult decision (Shmelev and Powell 2006, Lopolito, Nardone et al. 2011, Jensen and Wu 2016, Sadjadi and Karimi 2018). Rural communities are not the only ones with the mentioned traits. However, since they frequently lack the resources needed for resilience, they are more prone to experience the worst effects of bad decisions (Knickel, Redman et al. 2018, Ayala, Jurado et al. 2020).

The LEADER-CLLD (Liaison Entre Actions de Développement de l'Economie Rurale- Community-Led Local Development) approach, one of the cutting-edge techniques, has increasingly been employed throughout the European Union to promote the development of rural and regional areas. The primary feature of this approach is that it promotes the involvement of all stakeholders in the development process, which should not only be focused on economic activity but also social, cultural, environmental, and institutional factors (Sisto, Fernández-Portillo et al. 2022). Local Action Groups (LAGs) and the local development strategy are the main instruments of this approach (Sisto, Fernández-Portillo et al. 2022). The LAG oversees the formulation and carrying out of the development plan in light of the inherent characteristics of the proposed rural development model. By including representatives from a range of social groups in its governing bodies, it must accurately reflect the makeup of its region and represent the population's interest in the development processes being carried out. Besides, in the local development strategy, members of the LAG offer their opinions on how the region should develop over time (Bosworth, Annibal et al. 2016, Navarro, Woods et al. 2016).

One of the key challenges in this approach is the multidisciplinary nature of decision-making where stakeholders from all backgrounds may provide varied levels of input to the problems generally. As a result, proactive engagement of stakeholders via participatory methods including objective investigation of the competing priorities is highly crucial (Robinson, Burch et al. 2011). Multi-criteria decision-making (MCDM) methods are considered to be effective for managing and prioritising the conflicting criteria arising from the stakeholders' participatory process. For instance, Cuoghi and Leoneti (2019) reviewed the multicriteria decision analysis methods concerning Environmental Impact Assessment (EIA) as well as the energy sector. They reported that the group MCDA approach may be used to fill in gaps in the EIA methodologies and assist the public sector in studying complicated problems by breaking them into smaller units (Cuoghi and Leoneti 2019). In this regard, a combined tool such as a combination of multicriteria decision techniques and participatory methods is needed to be tailored for a specific problem to increase the consistency of the results and robustness of the study (Cuoghi and Leoneti 2019). Fernandez Portillo et al. (2019) highlighted the use of the Analytic Network Process (ANP) method for the prioritisation of strategies for rural development in Lagodekhi, Georgia under the LEADER approach.

There are many cases in the literature that demonstrate the effectiveness of participatory tools in the management of complex issues across the world. For instance, Lopolito and his co-workers (2011) have used a participatory approach to compare the available policy options in modelling a bio-refinery industry in rural areas. They explained that because of a lack of infrastructure, conflicting interests, a lack of knowledge, and a lack of human capital, promoting a new sector in rural regions is often tricky (Lopolito, Nardone et al. 2011). In this situation, public assistance is necessary to aid in the promotion of this cutting-edge industry (Lopolito, Nardone et al. 2011). Therefore, their empirical investigation was based on a simulation using the four specified policy instruments: financial support for bio-refineries, improving the profitability of biomass crops, technical advancement, and public education (Lopolito, Nardone et al. 2011). Each policy option's efficacy in terms of advancing bio-refineries has been taken into account (Lopolito, Nardone et al. 2011). Additionally, their investigation has looked into policies' pressure on unintended consequences (Lopolito, Nardone et al. 2011). Finally, the pressure and efficacy of different policies have been separated into direct and indirect impacts (Lopolito, Nardone et al. 2011). As a result, they noted that while subsidies and incentives to the profitability of speciality crops seem to have the most impacts on the growth of bio-refineries, technical innovation and informational options consistently outperform them. They attributed this to the fact that technological innovation and knowledge can take advantage of virtuous partners supporting the bio-refinery business with just minor adverse effects. Options based on subsidies and incentives for certain crops, however, place significant pressure on undesirable effects and may eventually lead to unsustainable patterns (Lopolito, Nardone et al. 2011).

As seen, participatory techniques can assist in integrating many viewpoints and expectations throughout the process to achieve the goal of accounting for the opposing aims, criteria, and expectations among various stakeholders. Besides, the sophistication of the analysis also rises when switching from a setting with a single decision-maker to one with numerous decision-makers as is the case with the LEADER-CLLD technique (Sisto, Fernández-Portillo et al. 2022). Therefore, the process of decision-making should encompass the participation part as well as the high-quality technique considering multiple criteria. Participatory methods make it possible to engage non-experts to play an active role and express their opinions, show their knowledge, and represent their values and preferences (van Asselt Marjolein and Rijkens-Klomp 2002). Besides, it may help overcome the drawbacks and failures of the "ungovernability" of complicated problems (Bijlsma, Bots et al. 2011) and address process uncertainty more actively, which would be advantageous for the legitimacy and execution of policies and their substantive quality.

The process of strategic planning for rural regions should consider conflicts between various interest groups, which typically have opposing goals, standards, and expectations. Unfortunately, while the LEADER-CLLD approach supported by the European Commission advocates the adoption of participative techniques, it does not explicitly suggest a process for prioritisation (Fernandez Portillo, Nekhay et al. 2019). Therefore, it would be required to establish a technique that first uses a participatory approach to identify the preferences and demands of rural stakeholders before allowing for their hierarchisation.

There are also numerous examples in the literature on the combined use of participatory methods and MCDM tools for optimising strategic rural planning and decision-making. For instance, Oddershede et al. (2007) have used the Analytic Hierarchy Process (AHP) to support the decisions for rural development in Chile. Baffoe (2019) has employed the AHP method to prioritise livelihood-related activities and interventions in terms of sustainable rural development in Fantekwa district, Ghana. Dereyurt and Gündüz (2020) stated that rural development includes a multiaxial structure that develops the socioeconomic structure, ensures non-agricultural economic diversification, and adopts a governance-based approach in the organisation and participation mechanisms rather than primarily on agricultural output. In this regard, they established an economic development model for the Güdül rural area in Turkey employing SWOT Analysis, AHP integrated method, and TOWS matrix with the goals of identifying the process' critical points, developing strategies intended for rural economic development within the axes valued by various stakeholders, and strengthening the participation mechanisms. Fabac and Zver (2011) have seen the development of rural areas through the lens of tourism support and used a modified SWOT-AHP method to analyse the available strategic options and to formulate the future tourism strategy of Gornje Međimurje, Croatia. Fernandez-Portillo et al. (2019) have used the ANP method to prioritise the rural development strategies under the rural development approach inspired by LEADER as this methodology does not propose a clear prioritisation method except for simple scoring methods and reported that ANP is better than less sophisticated methods in terms of adapting with LEADER approach. They also attributed this finding to the idea that ANP could efficiently compile the interactions between different dimensions of development (Fernandez Portillo, Nekhay et al. 2019) (Images 4.2 and 4.3).

Image 4.2 A purpose-built public water treatment plant for the village community in Assam.

Image 4.3 A light-weight wooden bridge serving as a high platform for recreational purposes on a wetland.

4.4 Role of local government and underlying power and responsibilities

The local government and institutions play essential roles in developing rural communities because they may foster an atmosphere that encourages innovation, creativity, and productivity among community members. Numerous studies have shown that local communities are best empowered by the local government due to their proximity to the location where the community reside and the authority for implementation. When local communities are given responsibilities in terms of the development and management of natural resources, this encourages them to propose solutions to problems and concerns connected to sustainable development (Ghai and Vivian 1995, Baland and Platteau 1996). Due to the virtue of local institutions and government being responsible for providing the required planning and administrative frameworks for rural development and enhancing the socioeconomic well-being of the disadvantaged and poor rural population, their roles are considered highly crucial (Curtis 1991, Douglas 2005). Local institutions such as primary, secondary or tertiary schools, Non-government Organisations (NGOs), community centres, clubs and hubs, play significant roles in the development of rural communities as resource mobilisers, service providers, program creators, supporters of resources, and integrators of resources (Khongsatjaviwat and Routray 2015).

While decentralised governance models are quite prevailing in many socialist countries for a long time, over the past decades, many new governments have introduced decentralisations in their governance models. In decentralisation governance, every administration strives to use decentralised strategies to boost

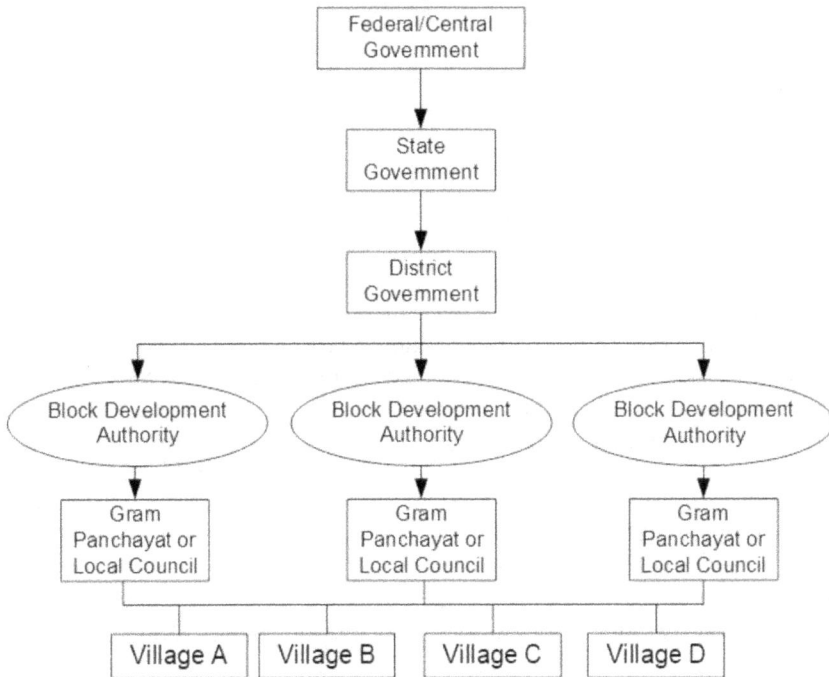

Figure 4.1 Decentralised governance model in India.

effectiveness and guarantee maximum local control. Anecdotally, many governments in Asia, including those in Japan, South Korea, China, and Vietnam, are increasingly relying on local governments to execute rural development schemes. Due to their firm understanding and proximity, local governments can carry out their duties and roll out the development schemes for the community effectively (Grindle 2007, Channa and Faguet 2016). Figure 4.1 shows the decentralised governance model in India introduced after independence.

As seen, decentralising governance models entices a greater likelihood of being accountable for executing policies at the grassroots levels. Federal or central governments especially in developing countries delegate the local governments substantial responsibilities with large portfolios across various sectors (Huther and Shah 1998). As a result, politicians, legal experts, and economists have started realising that the operation of local governments promotes transparency and accountability due to local populations being involved in constant monitoring of the progress of work and budgetary expenditures (Khongsatjaviwat and Routray 2015).

Decentralisation also appears to be well positioned given the rise of new public management in the 1980s because both new public management and decentralisation are opposed to heavily centralised governments (Khongsatjaviwat and

Routray 2015). The argument put out by decentralisation and new public management supporters is that carrying out governmental duties locally will increase local responsiveness and eradicate corruption. Many influential foreign funders have embraced the idea of decentralisation and provided poor nations with successful decentralisation records with financial support and counsel (Devarajan and Reinikka 2004).

The most significant theoretical justification for decentralisation is that it can increase government responsiveness and accountability to the governed. Real-world reformers' primary argument for their actions is to improve government. However, the literature has concentrated primarily on policy-relevant outcomes, such as public investment, fiscal deficits, and services for health and education (Faguet 2014). In this regard, Faguet (2014) reports that decades of study have revealed that decentralisation affects each of these outcomes differently in various nations and at different periods, preventing from drawing broader generalisations. Falkowski (2013) used the data of rural Poland to show the drawbacks of decentralisation combined with the LEADER program for rural development. Decentralisation has a varied effect on how public services are delivered. According to a meta-analysis of the literature conducted by Ghuman and Singh (2013), the impact of decentralisation on the provision of public services depends on several variables, including the design of the decentralisation policy, implementation bottlenecks, the dilution of the decentralisation model to account for dissenting stakeholder groups, including employees, and participatory governance.

Following are two examples, one from Nigeria and another from Thailand, highlighting the issues around decentralisation, governance and inefficiencies in the delivery of services (Image 4.4).

Image 4.4 Low-cost community-maintained public toilet connected with public-water supply system in a village in Assam.

4.4.1 Local government and rural development: A case study for Nigeria

Local government and rural development are interconnectedness terms and complementary to each other. Both terms signify the utmost importance in the field of public administration as they serve as the backbone of the growth trajectories of the countries, especially third-world countries. According to the Nigerian constitution, local governments are Nigeria's third tier of governmental institutions, after the federal and state governments (Ibietan 2010). The phrase "grassroots administration" is frequently used since it also illustrates where this kind of governance is located. However, due to a lack of resources and authority, the functioning of the local administration does not demonstrate high credibility and implementation records. Ironically, with overly centralised governance, Nigeria's constitutional provisions on decentralisation are somehow compromised (Ibietan 2010).

On the contrary, the concept of rural development and underlying objectives and techniques for the local government's efforts are clearly defined. The execution strategies of rural development are linked to the goals and targets of the specific programs. The contrast is due to the lack of maturity and transparency in the political system within the developing economies. Therefore, a crucial component of rural development is the deliberate blending of local (people's) efforts with the government to enhance socioeconomic circumstances and promote political engagement.

Since local government is the level of government closest to the people, a study conducted by Ibietan (2010) sought to define its function or position in rural development. It is reported that Nigeria's system and federal practice make it clear that local governments are weak and not well-positioned to fulfil their constitutional obligations, let alone the responsibilities they should play in rural development (Ibietan 2010). It is also added that self-help is a problem in rural development and several rural development strategies were examined (Ibietan 2010). In this regard, Ibietan (2010) highlighted seven-point recommendations as follows:

1 local governments' autonomy should be supported in a way that improves their ability to carry out their constitutional and auxiliary duties;
2 local governments should work to increase and sustain an income to implement initiatives and services that ensure excellent living for the general populace. This may lower the frequency of rural-urban drift;
3 local governments should adopt a more community-orientated philosophy so that required partnerships and engagement with local communities under their jurisdiction may aid rural development;
4 the revenue allocation process should be revised, aiming to improve the revenues to local governments since they are the closest layer of government to the community;

5 pragmatic political changes that might free up space for local governments and reduce the federal government's influence are urgently needed;
6 to teach and disseminate new farming techniques to farmers, local governments should take the initiative to employ agricultural extension workers and agents. This has the benefit of fostering essential adaptive agricultural techniques with increased output, which ensures food sufficiency and independence;
7 to energise rural development sustainably, it is essential to empower and link the institutions and groups that can help achieve rural development goals. The organisations and groups include cooperative movements, local/ town improvement unions, the National Directorate of Employment, and cooperative and agricultural banks.

4.4.2   *Local government and rural development: A case study for Tambon, Thailand*

Investigating the role of local government and local institutions for rural community development, Khongsatjaviwat and Routray (2015) highlighted some of the key issues being faced by Tambon Administrative Organisations (TAOs) which is a local implementation agency within the Phitsanulok province. Their findings asserted that due to a lack of appropriate funding, TAO was not able to function effectively. Consequently, many of the routine activities associated with local development could not be executed (Khongsatjaviwat and Routray 2015).

TAO in Phitsanulok province is composed of the legislative body of elected members under the Tambon Council (TC). The members are the representatives from local villages and the size of the executive body varies by the size of the village's jurisdictions. TAO is essentially the local government at the bottom of the governance ladder which is responsible for most developmental activities involving the local community at the grassroots level. While TAO's operations and decision-making are quite independent and highly discretionary, the scope and extent of development activities predominately rely on the annual grant money from the national government. Another avenue of TAO's budget comes from some of the local taxes but a portion of such income is fairly negligible. Thus the success of TAO's functioning and operation depends on their relationship with the national government which is considered highly inefficient in building and supporting local communities and meeting their needs and priorities in localised settings.

While the current investment patterns on local development projects in the TC support the development and maintenance of public infrastructure such as roads, water, and power networks, it excludes many of the pivotal livelihood-related projects that are necessary for thriving the community at local levels. Examples of some of such projects under TAO's risibility include:

• Establishment of a diesel fund to provide subsidies for the farmers and support agricultural activities for both local consumption and export opportunities

- Building new cultural centres and shopping strips for supporting local festivals such as Songkran Festival (Thai New Year), promoting local culture, food and heritage, and developing strong and collegial local communities
- Appropriate provision of infrastructure and management support for the cattle-raising community for promoting self-growth and livelihoods and thereby enticing the secondary income streams such as tourism for local businesses

Given the paucity of funding support, TAO is neither able to undertake any of the above projects nor has any financial independence for initiating new projects targeting the local communities. While the provisions in the decentralisation plan 2000 require the central government to support local institutions such as TAO with 35% revenue grants to support the basic minimum needs (BMN) for the community, TAO has not been receiving such a share of funds at all (Khongsatjaviwat and Routray 2015).

It can be concluded that under the current practice, TAO's increasing dependency on central government support for finance is hindering community development projects at the grassroots levels. While the paucity of funding is a chronic issue for TAO's operations, increased legislative and executive powers for collecting more taxes from citizens to meet the financial needs of the local projects are not considered a viable alternative. In summary, the current arrangement of decentralisation for prioritising local developmental projects is not quite effective and thus appropriate reforms along with raising of competencies and maturity need to be built in the overall governance structure.

4.5 Infrastructure for developing entrepreneurship and enterprise in rural and regional areas

In addition to the mere serviceability of the community, the role of rural infrastructure in promoting entrepreneurship is highly significant. As discussed in the above sections, the feasibility in rural infrastructure investment is not always a function of economic indicators such as return on investment or net present value. However, one of the underlying attributes of rural investment is the economic growth of the region including economic empowerment of the community at large. Entrepreneurship uptake and enterprising opportunities in rural and regional areas are some of the indirect and long-term measures of the performance of economic investments in the rural infrastructure system. Relevant policies for supporting rural entrepreneurship including building underlying capacities of the community for leveraging the opportunity need to be in place as a trade-off of the investment decisions and steady growth of the region.

4.5.1   Policies for rural entrepreneurship and enterprise

A wide variety of structures and large suits of programs and policies address the stimulation and support of various types of rural entrepreneurship. While some are national programs that include rural businesses alongside their counterparts in the rest of the country, others are focused mainly on rural businesses (North and Smallbone 2007). This is a result of the divergent historical roots of enterprise policies in various European nations, each of which reflects distinct cultural and ideological perspectives on the function of private industry in the economy and the function of state policy regarding the enterprise (North and Smallbone 2007). Additionally, it is linked to variations in governmental frameworks and the extent to which local and regional levels of government have participated in economic growth (North and Smallbone 2007).

When determining the scope of the policies, it is necessary to consider various aspects of entrepreneurship and company policies. In this respect, one can distinguish between more traditional enterprise support policies focused on the growth, survival, and competitiveness of businesses and entrepreneurship policies, which include the promotion of an entrepreneurial culture, entrepreneurship education, and policies to assist people through the early stages of starting a business (Stevenson and Lundstrom 2002). The particularities of the social, cultural, and political setting, as well as national-level policies regarding, for example, market openness, privatisation, and legal and tax regimes, are just a few of the numerous variables that might affect the amount of entrepreneurship within a rural economy (Glancey, McQuaid et al. 2000, North and Smallbone 2007).

Figure 4.2 shows some of the key focus areas for promoting rural entrepreneurship and enhancing rural regions' ability for entrepreneurship (North and Smallbone 2007):

- policies aimed at enhancing rural areas' potential for entrepreneurship, such as those that seek to change people's motives and attitudes towards it and offer opportunities for them to gain managerial and entrepreneurial capabilities via training and education;
- policies aimed at recruiting immigrants with entrepreneurship and goals or efforts aiming at raising the number of entrepreneurs from "underrepresented" populations, such as young people, women or people with disabilities and other marginal groups;
- policies aimed at assisting in the process of launching startups, such as pre-start-up guidance and evaluation of the marketing strategy as well as support for the many components of the business establishment;
- policies that offer general assistance to rural enterprises, such as those that offer guidance on various company operations such as strategic planning, advertising, import and exportation, supply chain and the utilisation of emerging as well as information technologies;

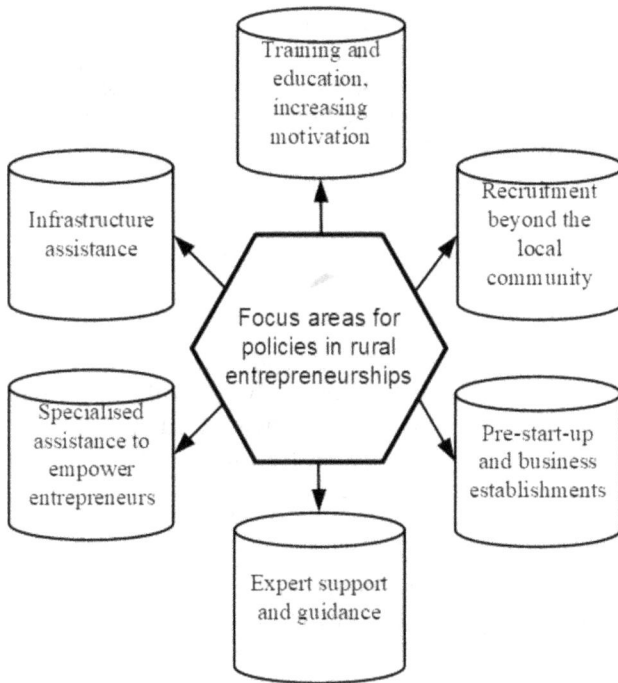

Figure 4.2 Policy consideration for promoting rural entrepreneurship.

- policies designed to offer specialised assistance to firms in certain industries, such as those assisting farmers to expand their enterprises into new "on-farm" or "off-farm" activities;
- entrepreneurship creation and growth in rural regions, not least as a way of overcoming some of the drawbacks of the periphery. For instance, the provision of a range of business facilities, including incubator units, financing in the transportation infrastructure required to facilitate market access, and investment in the Information and Communication Technologies (ICT) infrastructure to promote E-commerce.

The development of appropriate policies associated with financing entrepreneurship and company development initiatives is an important first step in informing the necessary need for underlying infrastructure. The policies need to be inclusive for promoting both social and economic development through entrepreneurship. Regional policies frequently include priorities and measures for new enterprise creation and small business development while generally adopting a reasonably holistic approach to the economic growth of specific locations, such as investments in hard and soft infrastructure (North and Smallbone 2007). These include programs that promote different community-based types of entrepreneurial engagement, such as social businesses and are focused on developing rural areas (particularly the EU's LEADER initiative) (North and Smallbone 2007). A good example for analysing as well as critically reviewing these policies could be found in the study of North and Smallbone (North and Smallbone 2007) in which ten case studies have been reviewed and explained how these policies applied in European countries.

4.5.2   *Entrepreneurial capacity of remote rural regions based on European case studies*

North and Smallbone (2007) discuss a variety of ways that policy interventions might help to encourage entrepreneurship and enterprise growth in peripheral rural regions after identifying several problems with the current policies in the case study locations. Given the variety of rural regions in Europe, as shown by North and Smallbone case study locations (North and Smallbone 2007), a common subject is the necessity for policies tailored to specific local conditions. In other words, specific rural environments have unique difficulties that need specific policy instruments that are attentive to and suited to those conditions. However, additional challenges are prevalent in various rural settings and can be addressed by more generalised strategies.

By the typology of rural policies, most rural policy interventions focus on enhancing the competitiveness of current rural businesses rather than on fostering rural regions' entrepreneurial capacity (North and Smallbone 2007). This is not entirely unexpected given that small and medium-sized enterprises-focused policies are more prevalent and that there are very few nations with explicit national policies encouraging entrepreneurship (North and Smallbone 2007). However, given the dearth of creative and dynamic businesses seen in many of the rural case studies in the mainstream literature (North and Smallbone 2007), the policy for supporting rural entrepreneurship certainly needs an overhaul, especially in developing economies. As a result, prospective entrepreneurial sources in remote rural locations would be easy to identify including nurturing and extending necessary support. According to the study of North and Smallbone (2007), these sources may include immigrants, young people, and a small group of significant entrepreneurs who consistently participate in many company initiatives and play a prominent role (animators).

4.5.3   *Entrepreneurship support by developing infrastructures*

Policy and the development of rural infrastructures that can support entrepreneurship are intertwined. However, the extent and nature of the influence of the policy could be affected by the degree to which private sector business services are developed, or they can satisfy the emerging entrepreneurs' needs in rural enterprises.

The necessity to fund the education and training infrastructures comes first, particularly in rural regions that have not progressed as much. A case study in rural Poland regions shows that the shortcomings of the existing educational infrastructures are a significant obstacle to a business's growth (North and Smallbone 2007). The relatively low skill and education levels of the rural workforce and potential entrepreneurs have a negative impact on the availability of business owners, the form and scale of enterprise development, particularly in technology and knowledge-based industries, and the efficiency and chances of success of new enterprises (North and Smallbone 2007). For this, it is necessary to make investments in secondary and tertiary education as well as in the training of entrepreneurs to improve their marketing expertise, capacity for creating business strategies, financial management, and innovation management calibre. It may also be necessary to reassess curriculum material and teaching strategies in secondary, further, and higher education during a transition or developing market economy (Gibb 1993).

With reforms to the physical and social infrastructure, it is also possible that the entrepreneurial ability of peripheral rural regions would be successfully developed. The periphery rural communities included in the "The Future of Europe's Rural Peripheries" research (Germany (Nordwestmecklenburg and Waldshut); Greece (Kilkis and the island of Lesvos); Poland (Zary and Bialystok); Portugal (Left Bank of the River Guadiana and the Oeste region); the UK (Cumbria, and Devon and Cornwall) are affected in varied degrees by this; nevertheless, the most disadvantaged areas are most immediately affected (North and Smallbone 2007). Therefore, to diversify Poland's rural economies by encouraging non-farming entrepreneurship, it was necessary to invest in rural areas' physical, technological, and social infrastructure (North and Smallbone 2007). This will create an environment favourable to the emergence of new businesses and multifunctional development and necessitates giving additional consideration to market institutions, banking systems, the delivery of education and training, and the adoption of new technology. It is further suggested that to promote innovation by local governments and communities, local self-governance should be promoted (Labrianidis 2004).

The lack of physical and social infrastructure in rural communities, whereby even parents advise their children to quit farming and seek jobs in the major metropolitan centres, is one of Greece's leading causes of the depopulation of peripheral rural regions (North and Smallbone 2007). Nevertheless, the retention of youngsters is a must for rural economic growth. Providing them with a source of money is an essential but insufficient requirement because they also need decent living circumstances, job possibilities, and social standing. To retain the kind of young people who are most likely to help these peripheral rural regions develop their entrepreneurial capacity, especially in the countries of southern Europe, medium-sized urban centres with the required physical and social infrastructure (roads, schools, provision of health facilities, etc.) are likely to be a requirement (North and Smallbone 2007).

4.6   Summary

After reviewing the broad contexts of rural infrastructure in the last three chapters, this chapter examined the factors associated with decision-making in infrastructure from the perspective of rural community development. This chapter is particularly important for understanding the challenges and opportunities associated with investment decisions in rural infrastructure and underlying rationalisation.

The first section of the chapter looks at how social protection, social welfare, and social insurance are some of the key considerations while community participation being is at the core of decision-making in infrastructure in rural and regional settings. Then the chapter discussion was expanded to some of the prevailing theories and practices of regional and rural policies where decentralisation of the governance structure was one of the key focus areas. Based on the evidence of two case studies, one from Africa and another one from Thailand, the pros and cons of decentralisation are discussed. While decentralisation models are considered to be highly effective for providing the necessary infrastructure for servicing the community and empowering bottom-up participation, how over-reliance on the funds of central or federal government hinders the functionality and effectiveness of the local governing bodies has been examined. The remedial course of action for the proper functioning of the local governing bodies along with the citizenry input for location-specific and need-based planning and development were highlighted.

In the last section of the chapter, the role of infrastructure in rural entrepreneurship and the promotion of social enterprises for the engagement of the broader community was discussed in detail. Policy support and capacity building among the community and the resultant benefits were highlighted as a potential measure of non-monetary rationalisation of rural infrastructure planning and development in contexts of rural and regional areas.

References

Ayala, L., A. Jurado and J. Pérez-Mayo (2020). "Multidimensional deprivation in heterogeneous rural areas: Spain after the economic crisis." Regional Studies **55**(5): 883–893.

Baffoe, G. (2019). "Exploring the utility of Analytic Hierarchy Process (AHP) in ranking livelihood activities for effective and sustainable rural development interventions in developing countries." Eval Program Plann **72**: 197–204.

Baland, J.-M. and J.-P. Platteau (1996). Halting degradation of natural resources: is there a role for rural communities?, Food & Agriculture Org.

Bijlsma, R. M., P. W. G. Bots, H. A. Wolters and A. Y. Hoekstra (2011). "An empirical analysis of stakeholders' influence on policy development: The role of uncertainty handling." Ecology and Society **16**(1): Article 51. http://www.ecologyandsociety.org/vol16/iss1/art51/

Bosworth, G., I. Annibal, T. Carroll, L. Price, J. Sellick and J. Shepherd (2016). "Empowering local action through Neo-endogenous development; the case of leader in England." Sociologia Ruralis **56**(3): 427–449.

Channa, A. and J.-P. Faguet (2016). "Decentralization of health and education in developing countries: a quality-adjusted review of the empirical literature." The World Bank Research Observer **31**(2): 199–241.

Conway, T., A. De Haan and A. Norton (2000). Social protection: New directions of donor agencies, London, Department for International Development.

Cuoghi, K. G. and A. B. Leoneti (2019). "A group MCDA method for aiding decision-making of complex problems in public sector: The case of Belo Monte Dam." Socio-Economic Planning Sciences **68**. https://doi.org/10.1016/j.seps.2018.04.002

Curtis, D. (1991). Beyond government: organisations for common benefit, London, Macmillan Education.

Devarajan, S. and R. Reinikka (2004). "Making services work for poor people." Journal of African Economies **13**(suppl_1): i142–i166.

Devereux, S. and R. Sabates-Wheeler (2004). Transformative social protection. IDS Working Paper 232. Brighton, Sussex BN1 9RE England Institute of Development Studies.

Douglas, D. J. A. (2005). "The restructuring of local government in rural regions: A rural development perspective." Journal of Rural Studies **21**(2): 231–246.

Economic and Social Council of United Nations (2001). Enhancing social protection and reducing vulnerability in a globalizing world: report/by the Secretary-General.

Fabac, R. and I. Zver (2011). "Applying the modified SWOT–AHP method to the tourism of Gornje Međimurje." Tourism and Hospitality Management **17**(2): 201–215.

Faguet, J.-P. (2014). "Decentralization and governance." World Development **53**: 2–13.

Falkowski, J. (2013). "Political accountability and governance in rural areas: Some evidence from the Pilot Programme LEADER+ in Poland." Journal of Rural Studies **32**: 70–79.

Fernandez Portillo, L. A., O. Nekhay and L. Estepa Mohedano (2019). "Use of the ANP methodology to prioritize rural development strategies under the LEADER approach in protected areas. The case of Lagodekhi, Georgia." Land Use Policy **88**: 104121. https://doi.org/10.1016/j.landusepol.2019.104121

Gentilini, U. and S. W. Omamo (2011). "Social protection 2.0: Exploring issues, evidence and debates in a globalizing world." Food Policy **36**(3): 329–340.

Ghai, D. P. and J. M. Vivian (1995). Grassroots environmental action: people's participation in sustainable development, Psychology Press.

Ghuman, B. S. and R. Singh (2013). "Decentralization and delivery of public services in Asia." Policy and Society **32**(1): 7–21.

Gibb, A. A. (1993). "Small business development in Central and Eastern Europe—Opportunity for a rethink?" Journal of Business Venturing **8**(6): 461–486.

Glancey, K., R. McQuaid and J. Campling (2000). Entrepreneurial economics, Springer.

Grindle, M. S. (2007). Going local, Princeton, Princeton University Press.

Huther, J. and A. Shah (1998). Applying a simple measure of good governance to the debate on fiscal decentralization, World Bank Publications.

Ibietan, J. (2010). "The role of local government in rural development issues" Knowledge Review **20**(2): 30–38.

ILO and WHO (2009). The social protection floor a joint crisis initiative of the UN Chief Executives Board for co-ordination on the social protection floor, Geneva, Switzerland.

International Labour Office (2001). Social security: A new consensus, Geneva, Switzerland, International Labour Organization.

International Labour Office (2011). Social protection floor for a fair and inclusive globalization: Social protection floor for a fair and inclusive globalization, Geneva, Switzerland, International Labour Office.

Jensen, O. and X. Wu (2016). "Embracing uncertainty in policy-making: The case of the water sector." Policy and Society **35**(2): 115–123.

Khongsatjaviwat, D. and J. K. Routray (2015). "Local government for rural development in Thailand." International Journal of Rural Management **11**(1): 3–24.

Knickel, K., M. Redman, I. Darnhofer, A. Ashkenazy, T. Calvão Chebach, S. Šūmane, T. Tisenkopfs, R. Zemeckis, V. Atkociuniene, M. Rivera, A. Strauss, L. S. Kristensen, S. Schiller, M. E. Koopmans and E. Rogge (2018). "Between aspirations and reality: Making farming, food systems and rural areas more resilient, sustainable and equitable." Journal of Rural Studies **59**: 197–210.

Labrianidis, L. (Ed.) (2004). The future of Europe's rural peripheries Burlington, VT 05401-5600, USA Ashgate Publishing Limited.

Lopolito, A., G. Nardone, M. Prosperi, R. Sisto and A. Stasi (2011). "Modeling the bio-refinery industry in rural areas: A participatory approach for policy options comparison." Ecological Economics **72**: 18–27.

Lovins, A. B. (1976). "Energy strategy: the road not taken." Foreign Affairs **55**: 65.

Midgley, J. (2012). "Social protection and social policy: key issues and debates." Journal of Policy Practice **11**(1–2): 8–24.

Navarro, F. A., M. Woods and E. Cejudo (2016). "The leader initiative has been a victim of its own success. The decline of the bottom-up approach in rural development programmes. The cases of Wales and Andalusia." Sociologia Ruralis **56**(2): 270–288.

North, D. and D. Smallbone (2007). "Developing entrepreneurship and enterprise in Europe's peripheral rural areas: Some issues facing policy-makers." European Planning Studies **14**(1): 41–60.

Oddershede, A., A. Arias and H. Cancino (2007). "Rural development decision support using the Analytic Hierarchy Process." Mathematical and Computer Modelling **46**(7–8): 1107–1114.

Ortiz, I. (2001). ADB's social protection framework. Social protection: New directions of donor agencies. T. Conway, A. de Haan and A. Norton: 40–63. Asian Development Bank.

Robinson, J. (2003). "Future subjunctive: backcasting as social learning." Futures **35**(8): 839–856.

Robinson, J., S. Burch, S. Talwar, M. O'Shea and M. Walsh (2011). "Envisioning sustainability: Recent progress in the use of participatory backcasting approaches for sustainability research." Technological Forecasting and Social Change **78**(5): 756–768.

Roelen, K. and S. Devereux (2013). Promoting inclusive social protection in the post-2015 framework. Policy briefing. Y. E. Azgad and H. Corbett. Brighton, UK, Institute of Development Studies.

Sabates-Wheeler, R. and S. Devereux (2007). "Social protection for transformation." IDS Bulletin **38**(3): 23–28.

Sabates-Wheeler, R. and S. Devereux (2011). Transforming livelihoods for resilient futures: How to facilitate graduation in social protection, Brighton, Future Agricultures Consortium, Institute of Development Studies.

Sadjadi, S. J. and M. Karimi (2018). "Best-worst multi-criteria decision-making method: A robust approach." Decision Science Letters **7**(4): 323–340. doi: 10.5267/j.dsl.2018.3.003

Şahin Dereyurt, B. and E. Gündüz (2020). "Assessment of the rural economic structure of Güdül town (Ankara) by quantified SWOT analysis." Iconarp International Journal of Architecture and Planning **8**(2): 672–702.

Sampson, T. and J. Drolet (2016). Key concepts and definitions of social protection, social development, and related terms. Social development and social work perspectives on social protection. J. Drolet. New York, USA, Routledge.

Shmelev, S. E. and J. R. Powell (2006). "Ecological–economic modelling for strategic regional waste management systems." Ecological Economics **59**(1): 115–130.

Sisto, R., L. A. Fernández-Portillo, M. Yazdani, L. Estepa-Mohedano and A. E. Torkayesh (2022). "Strategic planning of rural areas: Integrating participatory backcasting and multiple criteria decision analysis tools." Socio-Economic Planning Sciences **82**. https://doi.org/10.1016/j.seps.2022.101248

Sisto, R., A. Lopolito and M. van Vliet (2018). "Stakeholder participation in planning rural development strategies: Using backcasting to support Local Action Groups in complying with CLLD requirements." Land Use Policy **70**: 442–450.

Standing, G. (2007). "Social protection." Development in Practice **17**(4–5): 511–522.

Stevenson, L. and A. Lundstrom (2002). Entrepreneurship policy-making: Frameworks, approaches and performance measures. World Conference of the International Council for Small Business, San Juan, Puerto Rico.

Terluin, I. J. (2003). "Differences in economic development in rural regions of advanced countries: an overview and critical analysis of theories." Journal of Rural Studies **19**(3): 327–344.

van Asselt Marjolein, B. A. and N. Rijkens-Klomp (2002). "A look in the mirror: reflection on participation in Integrated Assessment from a methodological perspective." Global Environmental Change **12**(3): 167–184.

van Vliet, M. and K. Kok (2013). "Combining backcasting and exploratory scenarios to develop robust water strategies in face of uncertain futures." Mitigation and Adaptation Strategies for Global Change **20**(1): 43–74.

Wiesinger, G. (2007). "The importance of social capital in rural development, networking and decision-making in rural areas." Journal of Alpine Research | Revue de géographie alpine (95-4): 43–56. https://doi.org/10.4000/rga.354

5 Sustainability and infrastructure in rural and regional contexts

5.1 Introduction

Sustainability is a broad term and it usually encompasses a balancing act between economic, environmental, and social outcomes in most projects or operations. However, sustainability in rural settings refers to the actions with a direct impact on people's lives across every facet of human habitats. As over 40% global population still lives in rural areas, the scale and effects of actions on rural people across all the dimensions of sustainability are significantly high. Due to the virtue of rural communities being poor and underprivileged and with the increasing trend of the rapid urbanisation of the rural population around the globe, the demand for basic amenities and services is quite high. The lack of basic amenities in rural regions is one of the main factors pushing the rural population to urban centres across many countries. However, in recent years, with the increasing effort in rural investments, there is an unprecedented opportunity to increase the sustainability drive including bringing innovations for improving sustainability outcomes across the projects. This will eventually result in stemming the flow of urban migration, leading to a much-needed sustained share of rural and urban communities on the planet (Doloi 2017, Doloi 2023).

Sustainability efforts of the past three decades have certainly shown some positive outcomes, yet the bulk of the focus is on the urban sector projects. As the 40% rural population is confined to a few handful of areas such as African sub-continent, Indian-subcontinents, and Asian countries including China, significant work is needed in these regions towards improviement in economic, social, and ecological environment and improving people's lives. Among many available measures, United Nation's 17 SDGs are considered to be highly relevant for measuring improvement in people's lives holistically and objectively. However one of the key questions that remained unanswered is how these 17 SDGs are implemented in the context-specific environment for achieving sustainable development outcomes in rural projects. This sustainability chapter focuses on the argument of rural sustainability including highlighting the development and implementation insights in this context.

DOI: 10.1201/9781032622323-5

5.2 Defining sustainability and sustainable development

Basiago (1998) defines sustainability as the ability to maintain entities or a process over time (Basiago 1998, Mensah and Ricart Casadevall 2019). However, in the context of development, this word is being used as a concept that implies improvement and maintaining the economy, environment, and society for the development of mankind (Mensah and Ricart Casadevall 2019). In this regard, the two words sustainability and sustainable development are being used interchangeably in the literature. Sustainable development or sustainability has become a very reputable development pattern that can be found in many contexts (Mensah and Ricart Casadevall 2019). However, questions still arise about the true meaning of sustainability and its implications for relevant development theories and practices (Montaldo 2013, Shahzalal and Hassan 2019).

Conceptually, sustainable development goes back more than 200 years ago in the early 1800s when economists were debating the sufficiency of natural resources for the burgeoning population growth (Dixon and Fallon 1989). Malthus (1986), an economist, framed the basic idea of sustainability being an environmentalism where the reduction of adverse environmental impact is considered as the sole intent in all human actions. Following a similar ideology, containing the growth rate of the population and reducing the demand for depleting natural resources were considered some of the key measures in sustainability being accepted widely (Basiago 1998). However, with increasing concerns about the renewability of natural resources and depleting resources, sustainability-based discussion has become quite prominent, especially among the stakeholders involved in the project development and operations. Over time, the Brundtland Commission report provided one of the most accepted definitions of sustainability within the literature targeting stakeholders, researchers, and policymakers involved in sustainable development practices in projects (Brundtland Commission 1987). According to the United Nations' Brundtland Commission, sustainability (or sustainable development) could be defined as "meeting the needs of the present without compromising the ability of future generations to meet their own needs" (Brundtland Commission 1987).

5.3 Sustainability dimensions and underlying considerations

It seems that the idea of sustainability is continuing to influence the future of development science (Infrastructure Australia 2021). The principle of sustainability is all about ensuring the necessary provisions of goods and services for a thriving community without exerting adverse impacts on economic, social, or ecological aspects over the lifecycle phases of project development and operations. Therefore, the best options to choose are those not only are they environmentally friendly, viable in terms of economy, and at the service of the demands of society but also they are socially and economically equitable and socially and environmentally bearable (Porter and Van der Linde 1995). Therefore, three interconnected dimensions of sustainability in the rural infrastructure can be visualised in Figure 5.1.

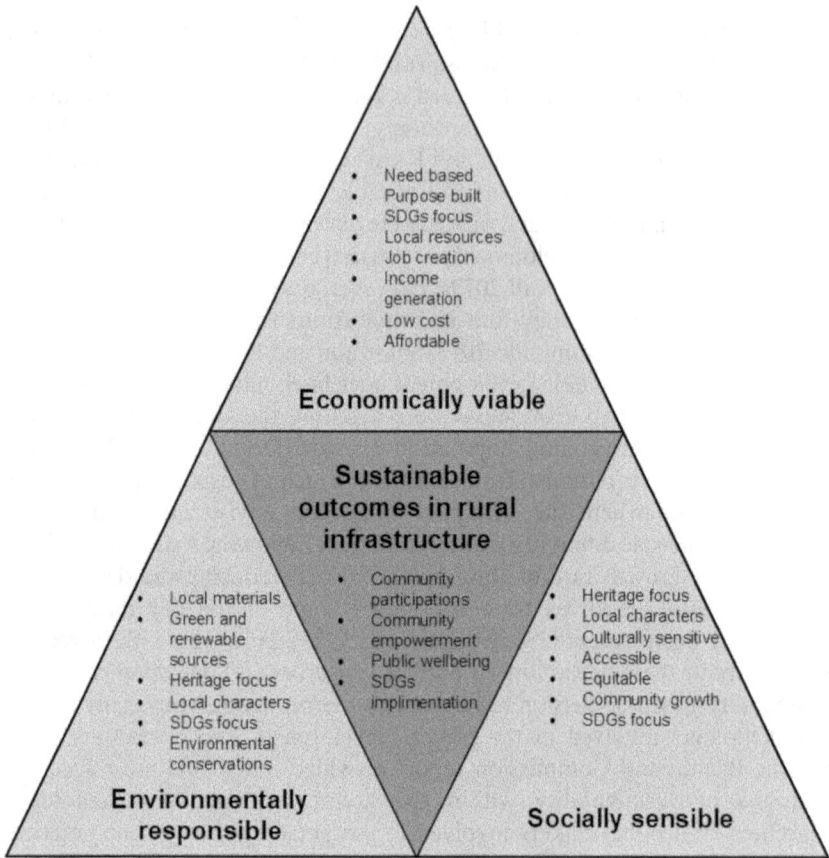

Figure 5.1 Sustainability dimensions in rural infrastructure.

While rural development ideas need to be economically viable, ecologically or environmentally responsible, and socially sensible, these dimensions are not necessarily to be mutually inclusive. Meaning, that the individual dimension needs to rationalise in its own right and when it comes to rural infrastructure, the threshold values of the underlying indicators under each dimension could be different from urban or semi-urban settings. In other words, the demarche of sustainable development in rural infrastructure needs to emphasise a pathway for achieving the interconnectedness of the social, economic, and environmental factors that result in the necessary transformation at the societal or community levels (Mensah and Ricart Casadevall 2019). Table 5.1 depicts the underlying considerations of all three sustainability dimensions from the perspective of rural infrastructure development. The dimensions are briefly discussed below.

Table 5.1 Sustainability considerations in rural settings

Sustainability considerations in rural settings		
Economically viable	*Socially sensible*	*Environmentally responsible*
• Economic viability of the infrastructure projects in rural settings is achieved through need-based infrastructure support that supports rural community for activities such as agriculture, community market, welfare centres, etc. • Such infrastructure facilitates community for performing their daily activities, engaging in jobs that earn income, support education and family welfare including growth of social enterprises and local entrepreneurships. • Planning, design, and development of rural infrastructure system must support the economic development of the community at large. Some of the key measures of economic development of the community include rise in income, increase in agricultural produces and export volumes, increase in literary rate, etc.	• Social sensitivity in rural settings is the functions of the preservations of the community heritage, social norms and values, cultural sensitivity and characteristics, equity, equality and accessibility to basic amenities for leading a comfortable life leading to social growth and healthy community. • Infrastructure systems which are predominately associated with social or community-based activities, must be designed and developed reflecting the heritage, culture and social norms so that it offers a conducive environment for the end-users and support the intended objectives in entirety. • Relative importance and significance of the social infrastructure need to be assessed and prioritised accordingly.	• Environmentally responsible infrastructure systems need to reflect on the localised context with use of local materials, local workmanship, and supporting engagement of the social enterprises across the project development lifecycle. • Project development lifecycle encompasses the project processes such as planning, initiation, construction or execution, operations, and maintenance. • Leading concepts such as renewable technology, circular economy, green energy, environmentally friendly materials including conservation of cultural heritage and social norms need to be incorporated in the project.

5.3.1 *Society and social sensitivity*

In the early years of research about the sustainability issue, many researchers' foci were centred around the economy and environment while social sustainability was left out of the radar in most considerations. The reason behind this negligence could be the fact that social sustainability is the least quantifiable among the pillars of sustainability. Some researchers used the theory of social organisations to define social sustainability. For instance, some researchers define it in very basic terms as "a system of social organisations that alleviates poverty" (Basiago 1998) and some such as Ruttan (1991) define it more fundamentally. Ruttan (1991) defines social sustainability as something that provides a series of connections between social conditions and the decay of the environment. However, there are still major debates on this theory of social organisations as some believe that environmental sustainability is needed before economic growth and the alleviation of poverty while others opine that economic growth and poverty alleviation are prerequisites of environmental sustainability (Basiago 1998). In more inclusive terms, social sustainability could be defined as an ideal status in which the interaction between the economy, environment and society brings about intergenerational equality and well-being (Ross 2013). In other words, social sustainability could be referred to as a kind of egalitarianism in well-being and quality of life between sociocultural groups over time and from one generation to the next.

As seen in Table 5.1, social sustainability in rural settings is more about functions such as preservation of the community heritage, social norms and values, maintaining cultural sensitivity and characteristics, equity, equality and accessibility to basic amenities for leading a comfortable life leading to social growth and healthy community. Infrastructure systems that facilitate social or community-based activities need to be planned and designed reflecting the heritage, culture and social norms so that it offers a conducive environment for the end-users and supports the intended objectives in entirety. While planning the social infrastructure, the relative importance and significance of the social infrastructure need to be assessed and prioritised based on its usages, needs, and requirements concerning the community concerned.

5.3.2　*Economy and economic viability*

The concept of economic sustainability, which stems from Hicks' study (1975), infers that a production process should meet the present consumption needs without compromising the future's prospects (Basiago 1998). Hicks defines income as an amount of money that a person can use during a period while having a good amount at the end of the period (Hicks 1975). In this regard, the economic sustainability of the rural infrastructure is a measure of how much income is being generated by the communities by engaging in economy-based activities.

The economic viability of the infrastructure projects in rural settings is achieved through need-based infrastructure support that supports rural communities for activities such as agriculture, community markets, welfare centres, etc. Such infrastructure facilitates the community to perform their daily activities, engaging in jobs that earn income, support education and family welfare including the growth of social enterprises and local entrepreneurship.

Image 5.1 Rural road with low-cost concrete pavers blocks surfaces with natural earth drains on the sides.

While planning the infrastructure for economic growth of the rural regions, the use of natural resources and underlying environmental impact are important to consider. In the past, economists used to assume that natural resources were unlimited and their emphasis was on the efficient allocation of resources (Basiago 1998). They also used to believe that the depletion of natural resources would be made up due to the technological advancements due to the fast-paced economic growth (Basiago 1998). However, the resources are not infinite and the growth rate of the economy and consumption also vary by the geographic and demographic compositions. Thus to achieve economic sustainability in rural settings, an integrated approach combining natural, social, and human capital-related factors needs to be considered (Goodland and Daly 1996). Planning, design, and development of rural infrastructure system must support the integrated development of natural, social, and human capital of the community at large (Image 5.1).

5.3.3 Environmental responsibility

Goodland (1995) has tried to distinguish the definition of sustainability social from social and economic sustainability. Although, a strong linkage exists between these three aspects of sustainability, providing a distinctive definition of environmental responsibility while creating facilities that offer goods and services for the common well-being of the community is important (Goodland 1995). Some of the earlier views of economists that unlimited resources exist for supporting the economy and social welfare of the population have been recalibrated over time regarding the depletion of natural resources and ecological imbalance. In this regard, Goodland (1995) defined environmental responsibility being a practice of allocation as "the relative division of the resource flow among alternative product uses – how much goes to the production of cars, to shoes, to teapots, etc.", distribution as "the relative division of the resource flow, as embodied in final goods and services, among alternative people", and scale as "the physical volume of the throughput, the flow of matter-energy from the environment as low-entropy raw materials, and back to the environment as high-entropy wastes". In addition, Goodland (1995) stated that the recognition of the deficiency of the market as a disruptive mechanism when the concern is about natural resources would also be necessary. Goodland (1995) finally defines environmental sustainability as a branch that focuses on the physical input from renewable and exhaustible natural resource bases into the process of production while emphasising the life support systems of the environment. Environmentally responsible infrastructure systems need to reflect on the localised context with the use of local materials, local workmanship, and supporting engagement of the social enterprises across the project development lifecycle. Leading concepts such as renewable technology, circular economy, green energy, and environmentally friendly materials including conservation of cultural heritage and social norms need to be incorporated into the project.

5.4 Measures of sustainable community

Infrastructure sustainability and sustainable community go hand in hand. The actions in developing sustainable rural infrastructure as discussed in the above section result in the creation of a sustainable rural community with tangible outcomes. Figure 5.2 depicts the fundamental measures of sustainable community across four key indicators, improvement in *well-being and quality of life*, exercising *fairness* across all sections of the community, and realising *inclusivity* and visibility in *social stability*. These indicators are further discussed below.

Figure 5.2 Fundamental measures of sustainable community.

5.4.1 Well-being and quality of life

There are wide-ranging examples in the literature explaining the role of infrastructures in the health and well-being of the community leading to better quality of life. The impacts of sustainable interventions in any community setting could be either positive or negative and therefore the exposed community could either enjoy the enhancement or suffer from the worsening of the well-being condition or the quality of life (Josa and Aguado 2019). For instance, as far as roads are important transport infrastructure for supporting the economy and mobility within the society, such infrastructure could also have an adverse impact by affecting the tranquillity and healthy recreational outdoor activities in the neighbourhoods. Investigating the role of transport infrastructure in psychological and physiological well-being, Conceicao et al. (2022) reported that congestion and delay are two key indicators of transport infrastructures linked to psychophysiological distress among commuters. Gross (2007) revealed that the well-being and health of a community are directly proportionate to the environmental management plans in the neighbourhood (Gross 2007). Indeed, there are other definitions and indicators for the community's well-being in the literature. Gross also stated that the ability to adapt to changes and adjust is a strong indicator of the healthiness of a community (Gross 2007).

5.4.2 Fairness

One of the outcomes that sustainability drives is the practice of fairness within the community. Gradoni and Murphy (2018) defined fairness in three components within a social setting, namely environmental justice, global justice, and intergenerational justice. In this categorisation, both global and intergenerational justice address the distribution of fairness in economy, policy, location, and geographic distribution, the time between locations including ease of logistics in terms of sending and receiving goods and services, etc. They further explained how important the state of environmental justice would be for both instrumental and non-instrumental reasons. The meanings of justice could range from simple issues such as rightfulness and fairness for accessing the needs and requirements (Macquarie 1996) to the more complex notions of equality or egalitarian distribution of benefits (Gross 2007). The famous law philosopher, Herbert Lionel Adolphus Hart, defined justice as "maintaining or restoring a balance or proportion" and stated that the two words of justice and fairness can be used interchangeably (Hart, Raz et al. 1961). Another prominent political philosopher, John Rawls, introduced two principles for justice by using an imaginary device he calls the original position or veil of ignorance, which is based upon the assumption that no one knows his social status, or his share in the distribution of intrinsic assets such as intelligence or strength (Rawls 1971). Thus, this veil of ignorance will lead to fair principles for everyone (Rawls 1971). His principles for justice include (Rawls 1971): (1) "Each person is to have an equal right to the most extensive total system of equal basic liberties compatible with a similar system of liberty for all"; (2) "Social and economic inequalities are to be arranged so that they are both: (a) to the greatest benefit of the least advantaged, consistent with the just savings principle, and (b) attached to offices and positions open to all under conditions of fair equality of opportunity". Since then, however, there have been major debates about the theories of justice (Gross 2007). For instance, Schlosberg (2004) criticised Rawls' theory and stated that this theory neglects the socio-cultural and institutional conditions and is primarily concerned with the distribution of goods and benefits. The other two important theories of justice are procedural justice and distributive justice. The main concern in the procedural justice theory is the process of decision-making to pursue societal goals such as different types of justice (Manaster 1995) while the distribution of outcomes either public goods or public burdens is the centre of focus of distributive justice (Kuehn 2000). Another category of justice that is frequently referred to by political theorists and justice critics is social justice (Gross 2007). Dobson (1998) has connected sustainability and justice by categorising their relations in three ways, i.e. "the environment as something to be distributed; justice as functional for sustainability; and justice to the environment". He claims that all theories concerning justice are about the distribution of something and this something is nothing but "environmental goods and bads". Therefore, he concluded that the relationship between justice and sustainability is embedded in the heart of the environmental justice movement (Dobson 1998). Dobson discussed that preconditions may be needed to meet the social objectives of sustainability and concludes that one of these prerequisites is social justice, which therefore, has a functional relation with sustainability (Dobson 1998). Apart from the mentioned distribution of something (environment) and justice in that distribution, Dobson's theory states that the environment itself should also be a recipient of justice (Dobson 1998). However, he stated, "neither environmental sustainability nor social justice has determinate meanings, and this opens the way to legitimizing the pursuit of either of them, in terms of the other, in several ways". Kuehn (2000), on the other hand, stated that the concern of social justice is the overall well-functioning of the society (Gross 2007) and defines environmental justice as a combination of social justice and environmentalism (Gross 2007).

5.4.3   Inclusivity

Inclusivity refers to the seamless integration of development programmes by closely aligning the project objectives with the community's expectations. The strategies for development with a focus on the economy may not serve as best as the project proponents may be concerned with the growth of income while compromising other facets of human life. However, social factors such as well-being in its general terms, or inclusivity could be far more important especially where the income growth allowed or viewed positively about the other critical factors. One of the good ways of contemplation the issue of inclusivity could be the United Nations' Agenda 2030 (UN General Assembly 2015) in which the SDGs have been specified and a set of frameworks, goals, values, and tools have been introduced by which a delivery of a more sustainable and more inclusive results would be possible (UN General Assembly 2015). These goals, indeed, range from environmental, and economic sustainability to social sustainability and fortunately as a result new development goals have come into the spotlight (Dörffel and Schuhmann 2022). In other words, not all of the development goals are income-related as it is widely accepted that the income may not directly and by its intrinsic value (Dörffel and Schuhmann 2022), affect the human well-being, as defined by United Nations Economic Commission for Europe (ECE 2014) as "broad concept which is not confined to the utility derived from the consumption of goods and services but is also related to people's functioning and capabilities", but is rather a mediator to promote other factors (Dörffel and Schuhmann 2022).

To address the issue of inclusivity, it would be central to define the concept of inclusive development (Doloi 2018). One of the most recent definitions is that inclusivity not only should encompass a fair distribution but also should have "preferable development returns". In other words, the concepts of inclusive development and equality are interrelated though they are not the same (Dörffel and Schuhmann 2022). Chatterjee (2005) has defined inclusive development as a process of development by which poverty and social exclusion will be reduced and a broad range of participation will take place.

In general, the definition of inclusive development has been contested in four distinguished ways (Pouw and Gupta 2017). In situations in which policymakers or scholars have used this term, inclusive development could mean from de-growth to levels far broader than growth. In this context, inclusive development could mean a development far broader than growth, or even a steady state of degrowth (Chatterjee 2005). Sometimes inclusive development could denote sharing the growth as well as increasing growth. As an example of this case, Ali and Son (2007) have stated that the availability of opportunities as well as how these opportunities are going to be distributed are the factors in assessing the inclusiveness of growth. Some scholars and policymakers, however, have defined the term as inclusive wealth. For instance, in the "Inclusive Wealth Report 2012: Measuring Progress toward Sustainability" report (UNU-IHDP 2012) it is stated that "the inclusive wealth framework moves away from the arbitrary notion of needs and redefines the objective of sustainable development as a discounted flow of utility which, in this case, is consumption".

The second way that the term inclusive development has been debated is where the concern was more focused on the development. This stems from the fact that development for some researchers and stakeholders could mean an increase in gross domestic product while for some others it could be as broad as the inclusion of social human needs (Costanza, Fisher et al. 2007), poverty reduction and well-being enhancement (Costanza, Fisher et al. 2007), financial inclusion (Chibba 2009), etc.

5.4.4   *Social stability*

Social stability refers to the conditions such as calmness, harmony, empathy, and mutual respect in society. Syme and Nancarrow (2001) used Pepperdine's (2001) indicators to highlight social stability and summarised the key ingredients needed for long-term stability in the context of rural communities. These ingredients include cohesion, mindedness, neighbourliness, acceptance of different viewpoints, community support groups, and conducive communication networks.

Increased social stability is linked to reduced poverty as well as improved income equality (Gurara, Klyuev et al. 2018). The reason behind this relation is that social stability allows us to focus and address the challenges of vulnerable people (GIHUB 2019) and make markets, services, and opportunities more accessible. In this regard, Estache and Garsous (2012) explained that social stability can augment the income of people and enhance productivity. In a research by Fan et al. (2002) on the growth, inequality, and poverty in the rural areas of China in 2002, it is reported that each 10,000 Chinese Yuan could lead to the lifting out of poverty of 3.2 persons. Investigating the effect of roads, irrigation facilities, and electricity on the community, Ali and Pernia (2003) asserted significant positive correlations between the infrastructures, poverty reduction, and social stability. These benefits of social stability in a well-governed community are in line with the first, second, eighth, and tenth (SDGs 1, 2, 8, and 10) goal of sustainable development goals (GIHUB 2019). Another strong relationship is demonstrated between inclusive infrastructures and social stability and equity (GIHUB 2019) which stems from the fact that these infrastructures facilitate the egalitarian distribution of benefits and reduce the divisions in the society (Lange, Wodon et al. 2018). This not only addressed the 16th goal (SDG 16) of sustainable development which is related to peace, justice, and strong institutions, but also it is in line with the eighth and tenth goals (SDG 8 and SDG 10). Inclusivity, infrastructure provisions, and underlying functionality for achieving SDGs are further discussed in the following section.

5.5   Infrastructure functionality for achieving SDGs

The efficacy and outcomes of rural infrastructure depend on how the functionality of the infrastructure system supports the community in performing context-specific activities. Such activities not only enable productivity growth in the rural community but also facilitate the vast majority of the people towards building a sustained and resilient society at large. While the SDGs are usually good indicators of how the infrastructure systems contribute towards realising the goals, close alignment of the underlying functions is important first step in planning sustainable infrastructure and supporting the growth in the community in rural settings.

Infrastructure systems are the backbone of every society, thus the UN's 17 SDGs act like a lifeline in supporting and flourishing rural communities. The SDGs are the collective forces for providing the necessary services and empowering with self-sufficiencies within the community. A purpose-built infrastructure system is also a great source of mutual trust and transparency in governance for improving the quality of life among rural communities. In this section, we aim to connect the infrastructure needs concerning the 17 sustainability goals so that a detailed understanding of the interdependent infrastructure network can be visualised (Thacker, Adshead et al. 2019).

Table 5.2 depicts the functionalities of the infrastructure system concerning the 17 SDGs and underlying infrastructure support. As seen, while economic growth has been considered important in traditional measures of success in infrastructure investments, the success in rural settings is much more rounded towards contextual implementation of the SDGs. Among many different sectors of infrastructure, transport infrastructure is usually considered to be highly significant in economic growth and development. Due to a large rural population vastly relying on an agricultural-based economy globally, transport infrastructure serves as a lifeline, facilitating functions such as accessibility, and movements of goods, services and people (Fourie 2006, Singh, Mathiassen et al. 2010). In many developing countries including Sub-Saharan Africa and India, the population engaged in agriculture-based activities is over 50% of the total population and it's usually the single most critical sector as far as the employment opportunity is concerned. In rural areas where usually agricultural lands are available, population distribution is uneven and thus reliance of the farming community on transport infrastructure is quite high. Thus, the importance of transport infrastructure for mobilising goods, services and people is quite significant which eventually pushes the demand for road and transport infrastructure across the rural areas. A lack of appropriate transport infrastructure connecting rural communities would potentially trigger in lack of opportunities for income generation which would result in increased poverty and poorer communities. Along with the transport infrastructure, appropriate shelter and storage infrastructure are also important for supporting economic activities within the farming community. Referring to the SDG goals, Goal 1: No Poverty, Goal 2: Zero Hunger, Goal 10: Reduce Inequality, and Goal 12: Responsible consumption and production are intricately linked to the integrated transport, housing and industrial infrastructure systems. While Goals 1 and 2 are directly linked to the efficacies of the agriculture-based economy in rural settings, Goals 10 and 12 can be targetted as a resulting phenomenon of efficient multi-level governance, education and skills development supported by relevant policies and infrastructure provisions (Thacker, Adshead et al. 2019) (Image 5.2).

Table 5.2 Infrastructure functionality across 17 SDGs

SDGs	Infrastructure functionality
	• Inclusive and integrated infrastructure system • Increase access to goods and services, marketing and local trades • Increase access to basic needs, shelter and food • Increase access to education, learning and skills development • Provide equal access and support for everyone
	• Accessible health infrastructure such as local dispensary, tele-health services, emergency medicine including ambulance services and seamless operations • Affordable High-speed Internet services • Computer facilities through local health centres • Infrastructure for producing, storing and distributing potable water • Eco-friendly sanitation and waste-disposal systems
	• Location-based infrastructure for primary, secondary, tertiary and continuing education for adult population • Vocational training and skills centre with context and demand-based support services • Access to education, skills development and training for all segments of the community • Instructure support for reskilling opportunities for all • Institutional partnership networks and mutual education
	• Infrastructure for clean, green and affordable energy • Production plants and distribution network for agricultural-based production, bulk-handling and storage facilities • Local trade centres, supporting facilities such as banking, postoffice, or other micro-financing agencies for social enterprises, non-governmental organisations, and social groups • Local and regional connectivity between the villages and semi-urban centres for trading and seamless support for localised transport and supply-chain
	• Facilitation of the green, natural and vernacular construction materials, processes and technology • Local trades and workmanships including localised industry for local construction and supplies

(Continued)

Table 5.2 (Continued)

SDGs	Infrastructure functionality
	• Facilities for waste management, waste reduction, recycling, reuse and campaign for adopting circular economy • Support systems for human capital growth and development towards sustainable living
13 CLIMATE ACTION 15 LIFE ON LAND 14 LIFE BELOW WATER 16 PEACE, JUSTICE AND STRONG INSTITUTIONS	• Localised materials, resources, need-based production and consumptions • Appropriate provisions including retrofitting of the facilities for measures such as energy efficiency, rainwater harvesting, waste reduction, ethical behaviours, and lifestyle • Facilitation of the circular economy, green, and vernacular lifestyle • Social enterprises for supporting shared facilities, community-based support systems
9 INDUSTRY INNOVATION AND INFRASTRUCTURE 17 PARTNERSHIPS FOR THE GOALS	• Social enterprises for supporting shared facilities, community-based support systems for organic growth of the local industry with innovation and incubations • Organically grown and needs and demand-based partnerships among social enterprises and community groups including supply of local goods and services

Image 5.2 Rural road with low-cost bitumen surface on crushed-rock sub-base with natural earth drains on the sides.

The targets for good health and well-being (Goal 3) and clean water and sanitation (Goal 6) can only be met if provisions are made for relevant infrastructures in the plan (Thacker, Adshead et al. 2019). Accessible health infrastructure such as local dispensaries, telehealth services, and emergency medicine including ambulance services and operations need to be planned based on the underlying demographic and geological conditions of the target community. For functioning of the telehealth, information and telecommunication infrastructure including high-speed internet are crucial. Hardware such as computers with necessary accessories along with security, maintenance, and efficient operations need to be planned, procurement, and installed for improving the accessibility to the seamless healthcare to the community. All these provisions would support the implementation of Goal 3 reasonably well. Concerning Goal 6 which is clean water and sanitation, appropriate planning and implementation of infrastructure provisions such as water treatment and wastewater treatment plans, catchment and watershed management, flood protection and management as well as filtration, supply and distributions networks are highly crucial (Ruan, Wang et al. 2015). On the waste disposal front, eco-friendly sanitation and waste management systems should be installed by utilising local knowledge, skills, and resources.

In regards to quality education (Goal 4), location-based infrastructure for primary, secondary, and tertiary including continuing education for the adult population is a basic need for imparting quality education across all relevant sectors in the community. Need-based and demand-specific vocational education, training and skills development centres are mandatory for upskilling the community with quality education and provide every segment of society with a clear growth trajectory. Gender equality (Goal 5) is closely coupled with the quality, unbiased and skills-based education and training systems for all in the community. Institutional partnerships include careful integration of social enterprises for education and the development of relevant skills that can genuinely complement the growth across all sections of the community in a collegial manner.

Affordable and clean energy (Goal 7) is not only important for community access from equity and health and well-being perspectives but also crucial for minimising carbon footprints and stemming the unprecedented rate of rise in temperature globally. Thus state-of-the-art, modern, and up-to-date infrastructure facilities would be necessary for achieving affordable and clean energy targets and facilitating growth of the community enticing futurist opportunities in the development ladder. Decent work and economic growth (Goal 8) are closely linked with the economic benefits of the rural transportation infrastructure concerning market accessibility, household consumptions and income, growth in human indices and local GDP. Accessible production plants and distribution networks for agricultural products including bulk storage and handling facilities can exemplify the local trading and marketing opportunities among the rural community. For instance, because a vast majority of African farmers usually live on average five hours or more from the local trade centres and markets, yet easy access to the local transport networks coupled with affordable prices make doing some of the key economic activities possible and support towards Goal 8. Decent work and economic prosperity of the rural people are also dependent on the facilities such as functional and accessible local trade centres, associated facilities such as banking, post offices, or other micro-financing agencies for supporting social enterprises, non-governmental organisations, and social groups. Local and regional connectivity between the villages and semi-urban centres for trading and seamless support for localised transport and supply chain is another key consideration for creating job opportunities and economic growth (Thacker, Adshead et al. 2019).

In regards to Goal 11, the sustainable cities and community, facilitation of green, natural and vernacular construction materials, processes and technology are the key requirements to realise this goal. Local trades and workmanships including localised industry for local construction and supplies, facilities for waste management, waste reduction, recycling, reuse, and campaigns for adopting a circular economy are also some of the key ingredients for creating sustainable communities in rural settings. Sustainable cities and communities can grow only when human capital grows. Human capital growth is usually the function of underlying living indices with a minimum level of performance against the respective thresholds in the specific community contexts (Cloutier, El-Sayed et al. 2022). Relevant infrastructure and an integrated network for supporting such targets are crucial to realise the benefits in the long run.

The remaining six SDGs for integrated planning and development of rural infrastructure systems include Goal 9: Industry innovation and infrastructure, Goal 13: Climate action, Goal 14: Life below water, Goal 15: Life on land, Goal 16: Peace, Justice and strong industries, and Goal 17: Partnerships for the goals. Industry innovation in rural settings links to the use of localised materials and resources for need-based production and consumption. Appropriate provisions including retrofitting of the facilities for measures such as energy efficiency, rainwater harvesting, and waste reduction including innovative solutions for reducing carbon footprints and improving efficiencies in operations and maintenance are some of the key measures for driving productivity, promoting peace and justice, and building stronger industries. These actions would potentially contribute to meeting the target towards Goal 13, Goal 16, and Goal 17. The target for meeting Goals 13, 14, and 15 requires appropriate support systems and infrastructure provisions for promoting ethical behaviours and lifestyles among the rural community. Actions such as facilitation of the circular economy, and green and vernacular lifestyle are important especially as contributions towards Climate Actions and making the planet sustainable. Social enterprises for supporting shared facilities, and community-based support systems for organic growth of the local industry with innovation and incubations are highly crucial for contributing towards the target of Goals 9, 15, and 16. Organically grown needs and demand-based partnerships among social enterprises and community groups including the supply of local goods and services support the growth in industrial practices and thereby enhance the lives both on land (Goal 15) and below water (Goal 14).

There are however examples where many of the basic targets associated with the SDGs are not getting delivered due to lack of appropriate rural infrastructure provisions. Studying the impact of transport infrastructure on events such as pandemics or mental stresses in the community, Ruan et al. (2015) emphasised the necessity of need-based and location-specific design of infrastructure to manage such events adequately. They asserted that inappropriate design has the potential to exert a harmful impact on people's lives, resulting in breaching the goals of sustainability generally. Investigating the emissions of a few powerplants, Pfeiffer and his co-workers (2018) reported higher emissions, exceeding the allowable carbon emission limits as per the Paris Climate Agreement. Investing in the relationships between physical infrastructure and accessibility, Bajar and Rajeev (2016) reported some high dispersity and increasing inequality across 17 Indian states in regional settings. One of the underlying principles in planning and designing rural infrastructure systems is appropriate provision for the management of the interdependencies and collective contributions towards the meeting the sustainable outcomes. The ripple impacts of failure events from a single infrastructure into larger infrastructure networks hinder sustainable operations which eventually result in compromising the sustainable outcome by triggering factors such as inaccessibility, inequality, etc. (Hall, Otto et al. 2016). The importance of interdependencies and underlying interventions needs to be prioritised for achieving integrated sustainability goals within the rural community.

5.6   Processes of creating sustainable infrastructure in rural settings

While sustainability as a concept and underlying actions required for achieving the targets are relatively well-documented across vast literature, there is a lack of demonstration of how the processes of creating tangible sustainability outcomes work in rural settings. UN's SDGs set a strong foundation for sustainable drive, especially in rural locations. Yet, the implementation of SDGs and meeting the sustainability targets in context-specific conditions requires practical demonstration. Often, policy-makers and front-line professionals lack transferrable knowledge and skills on the implementation front. The remainder of this section focuses on clear articulation of the processes of creating sustainable infrastructure systems as a contributing force towards achieving SDGs targets in rural conditions.

Figure 5.3 illustrates some of the key processes in creating sustainable infrastructure systems in rural settings. As seen, on the horizontal timeline, there are three important phases, *decision points, process and actions,* and

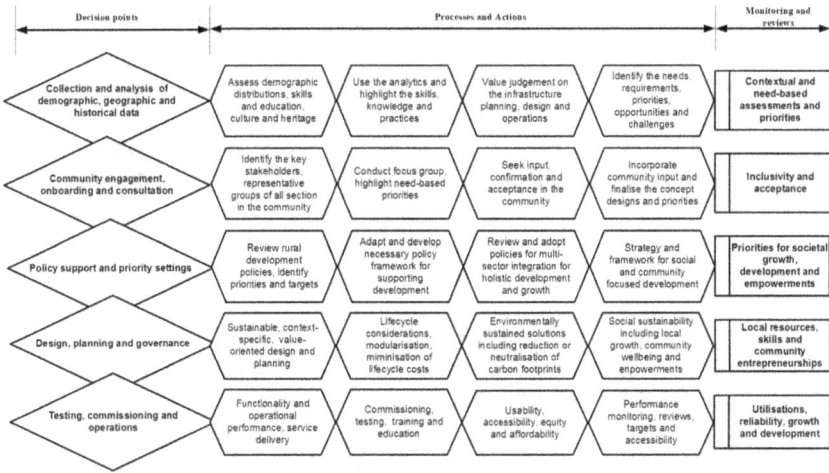

Figure 5.3 Processes in creating sustainable infrastructure systems in rural settings.

monitoring and reviews. Under each of these phases, there are a series of important indicators for guiding the processes in the creation of sustainable outcomes through infrastructure interventions. The significance of these indicators is discussed in the following sections.

As seen in the figure, under the first phase "decision points", five key indicators are listed vertically which should be viewed as progressive checkpoints in the sustainability ladder in the development. These indicators are further discussed about the "processes and actions" and "monitoring and reviews" as depicted in the horizontal timeline.

- *Collection and analysis of demographic, geographic and historical data:* The first action point is an assessment of demographic data based on a comprehensive field survey capturing skills, education, culture, heritage, etc. Such data is further processed using data analytics to highlight the salient characteristics of the entire community. As revealed by the data analytics, a value judgement is then undertaken for conceptual planning of infrastructure needs and the objectives are closely linked to the needs, requirements, and priorities including strategies for leveraging opportunities and challenges concerning the community concerned. The findings of these processes and actions are then closely monitored and reviewed ensuring meeting or exceeding the target and maintaining the sustained outcomes.
- *Community engagement, onboarding, and consultation:* Processes and actions include identification of key stakeholders inside and outside of the community, conducting focus group discussions to highlight need-based priorities, seeking acceptance within the broader community, refining and

fine-tuning with the community's input for ensuring complete inclusivity and acceptance of the proposed development propositions.

- *Policy support and priority settings:* These indicators include the processes of reviewing the rural development policies from a broader perspective for needs and priorities at the grassroots level, adopting and developing a necessary policy framework for supporting development, reviewing and adopting policies for multi-sector integration for holistic development and growth strategies, aligning social-focused and people-centric development planning. The efficacies of these processes need to be monitored and reviewed to ensure shifts in priorities if any, and quantitative measures of development and empowerment.
- *Design, planning, and governance:* These indicators include sustainable, context-specific, value-oriented design and planning, lifecycle considerations, modularisations aligning variability in demand and supply, environmentally sustained solutions including neutralisation of carbon footprints, social sustainability for promoting social stability, community welfare, fairness, growths, and development. Engagement and utilisation of local resources, and skills including social and community entrepreneurship are some of the tangible measures of these indicators in the development process.
- *Testing, commissioning, and operations:* The final set of indicators includes testing the functionality, operational performance and services being delivered, commissioning, training, training and imparting continuous education, usability, accessibility, equity and affordability, performance monitoring, reviews, and measure against performance thresholds resulting from the project. Reviewing the real-time operational data, performance on utilisations, reliability, growth and development need to be monitored, reported, and assessed in short, medium, and long terms (Image 5.3).

Image 5.3 A home-stay tourist lodge on a wetland for experiencing nature-based living and supporting non-agricultural local economy.

5.7 Urban-rural share, sustainability, and sustainable planet initiatives (SPI)

Over past decades, one of the key outcomes from the sustainability drive especially in the context environmental dimension is the reduction of emissions of greenhouse gases and contributing to the net-zero targets being set by various countries in the world. Man-made destructions of the natural ecosystem resulting from unsustainable land use, and lifestyles including over-consumption and production, are responsible for global warming and increasing surface temperature (IPCC 2023). Ironically, with the global population touching the 8.1 billion mark in 2024 and considering the rapid urbanisation of the 40% plus rural population who are still living in remote and underdeveloped conditions, especially in developing countries, it is hard to imagine that human activities against the nature are coming to an end in a foreseeable future. As reported in a recent IEA report "The top 1% of emitters globally each had carbon footprints of over 50 tonnes of CO2 in 2021, more than 1000 times greater than those of the bottom 1% of emitters" (IEA 2023). As per the report, there is a huge variation across income groups and it states that per capita energy consumption among the higher income earners is 11 times higher than the lower income earners. Thus, increasing carbon emission is linked to the wealth and lifestyles of the people predominately residing in urban areas across the countries.

Regarding the sustainability practices being put forward by the Author through the Smart Villages research programme, it has been contemplated that the 40% global population who still resides in rural regions (the so-called villagers) need not race against time to become urbanites by way of developmental transformation. The current global trend of closing the gap between the urban-rural divide is neither sustainable nor desirable (Doloi 2023). Both urban and rural communities have specific purposes and responsibilities in life. For instance, without farmers with farms in regional areas, the urbanites wouldn't get the food for survival. Similarly, cities are important business centres for supporting the economics of the countries. Among the global population, 60% urban and 40% rural is a meaningful and healthy divide, and efforts are needed to maintain this balance for a sustained future. Unfortunately, there is a clear lack of understanding, knowledge, skills, and awareness among the policy-makers, practitioners including other related stakeholders for upgrading these 40% of rural people with an alternative and context-based sustainable development model. The alternative sustainable development models need to integrate the vernacularisms of the rural community and build on thousands of years of local experience in leading healthy, nature-based sustained lifestyles.

Instead, the urban-centric knowledge and practices (the so-called brick-mortar development) are being extended in the name of modernising these 40% rural communities which not only violates every single principle of sustainable development in gross terms discussed throughout this book but also exerts adverse impacts on the climate significantly.

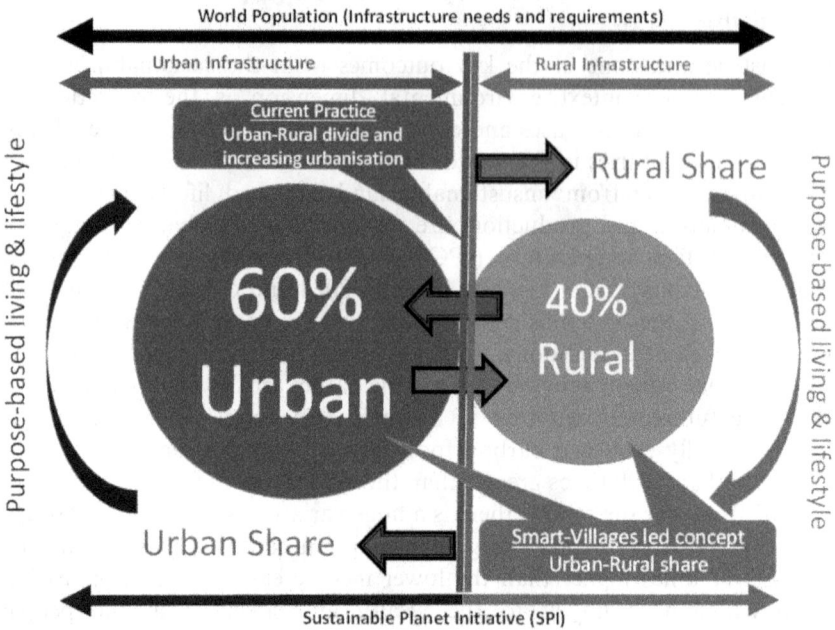

Figure 5.4 Urban-rural share and sustainable planet initiative (SPI).

Considering that the 40% rural population who are not only living in rural vernacular conditions but also leading a relatively low carbon, if not carbon neutral, lifestyle, alternative, and sustainable development models for modernising these groups of people on Earth is the need of the hour. Figure 5.4 illustrates a conceptual model for promoting rural development with a focus on "Urban-Rural Share" contrasting the traditional "Urban-Rural divide" and underlying efforts in closing the gap. The outcomes of such an alternative sustainable development model will contribute in a big way by containing the emissions of greenhouse gases and assist in meeting the net-zero targets not only nationally but perhaps at a global level which is being conceptualised as a sustainable planet initiative (SPI) and a green mission within the smart villages lab (SVL).

5.8 Summary

In this chapter, sustainability and infrastructure were discussed as a mutually inclusive deliverable in the context of rural and regional settings. With a quick review of sustainability and sustainable development from a historical perspective, three core sustainability dimensions in rural contexts have been highlighted. These dimensions revealed the interconnectivity of society and social sustainability, economic and economic viability, and environmental responsibility as a guiding principle of rural development.

As much as sustainability practices are necessary, so are the measures of a sustainable community through indicators such as well-being and quality of life, fairness, inclusivity, and social stability. These indicators are closely aligned to the UN's SDGs and necessary infrastructure provisions need to be in place as a key enabler to support the community. The discussion is then expanded on identifying the core infrastructure functions concerning the 17 SDGs so that a holistic understanding can be developed before deciding on infrastructure systems in specific geographical settings. Step-by-step processes and actions required for creating a sustainable infrastructure system are further established concerning a set of five key indicators. Tangible targets against each process are highlighted from monitoring and reviewing perspectives.

Focusing on an ideological viewpoint for maintaining "Urban-Rural share" among the global population, a clear position between urban and rural communities is established as a way of promoting the SPI. The need for alternative sustainable development models based on vernacularism, heritage, and values of the rural communities for promoting a low-carbon footprint, if not the carbon-neutral lifestyles among the 40% rural community is highlighted.

References

Ali, I. and E. M. Pernia (2003). "Infrastructure and poverty reduction-What is the connection?" Economics and Research Department (ERD) Policy Brief Series, Number 13, January, Asian Development Bank (15 pages).

Ali, I. and H. H. Son (2007). "Measuring inclusive growth." Asian Development Review **24**(1): 11–31.

Bajar, S. and M. Rajeev (2016). "The impact of infrastructure provisioning on inequality in India: does the level of development matter?" Journal of Comparative Asian Development **15**(1): 122–155.

Basiago, A. D. (1998). "Economic, social, and environmental sustainability in development theory and urban planning practice." Environmentalist **19**(2): 145–161.

Brundtland Commission (1987). Our common future (the Brundtland Commission report). Report of the World Commission on Environment and Development, United Nations.

Chatterjee, S. (2005). "Poverty reduction strategies—Lessons from the Asian and Pacific region on inclusive development." Asian Development Review **22**(01): 12–44.

Chibba, M. (2009). "Financial inclusion, poverty reduction and the Millennium Development Goals." The European Journal of Development Research **21**(2): 213–230.

Cloutier, S., S. El-Sayed, A. Ross and M. Weaver (2022). Linking sustainability and happiness: theoretical and applied perspectives. Switzerland, Springer Nature.

Conceição, M. A., M. M. Monteiro, D. Kasraian, P. E. W. van den Berg, S. Haustein, I. Alves, C. L. Azevedo and B. Miranda (2022). "The effect of transport infrastructure, congestion and reliability on mental wellbeing: a systematic review of empirical studies." Transport Reviews **43**(9): 1–39.

Costanza, R., B. Fisher, S. Ali, C. Beer, L. Bond, R. Boumans, N. L. Danigelis, J. Dickinson, C. Elliott, J. Farley, D. E. Gayer, L. M. Glenn, T. Hudspeth, D. Mahoney, L. McCahill, B. McIntosh, B. Reed, S. A. T. Rizvi, D. M. Rizzo, T. Simpatico and R. Snapp (2007). "Quality of life: An approach integrating opportunities, human needs, and subjective well-being." Ecological Economics **61**(2–3): 267–276.

Dixon, J. A. and L. A. Fallon (1989). "The concept of sustainability: Origins, extensions, and usefulness for policy." Society & Natural Resources 2(1): 73–84.

Dobson, A. (1998). Conclusion. Justice and the environment: conceptions of environmental sustainability and theories of distributive justice, Oxford University Press: 240–262.

Doloi, H. (2017). "Smart Villages Lab (SVL)." from https://smartvillageslab.msd. unimelb.edu.au/.

Doloi, H. (2018). "Community-centric model for evaluating social value in projects." Journal of Construction Engineering and Management 144(5). https://doi.org/10.1 061/(ASCE)CO.1943-7862.0001473

Doloi, H. (2023). Smart data-driven approach for developing resilient society. The 2nd International Conference on construction resources for environmentally sustainable technologies (Crest 2023). H. Hazarika. Fukuoka, Japan, Kyushu University: 10.

Dörffel, C. and S. Schuhmann (2022). "What is inclusive development? Introducing the multidimensional inclusiveness Index." Social Indicators Research 162(3): 1117–1148.

Ece, U. (2014). "Conference of European Statisticians recommendations on measuring sustainable development/United Nations Economic Commission for Europe; prepared in cooperation with the Organisation for Economic Co-operation and Development and the Statistical Office of the European Union (Eurostat)."

Estache, A. and G. Garsous (2012). "The impact of infrastructure on growth in developing countries." IFC Economics Notes. Note 1. Washington D.C.: International Financial Corporation.

Fan, S., L. Zhang and X. Zhang (2002). Growth, inequality, and poverty in rural China: The role of public investments, Intl Food Policy Res Inst.

Fourie, J. (2006). "Economic infrastructure: A review of definitions, theory and empirics." South African Journal of Economics 74(3): 530–556.

Gardoni, P. and C. Murphy (2018). "Society-based design: promoting societal well-being by designing sustainable and resilient infrastructure." Sustainable and Resilient Infrastructure 5(1–2): 4–19.

GIHUB (2019). Inclusive Infrastructure and Social Equity: Practical guidance for increasing the positive social outcomes of large infrastructure projects. Victoria, Australia, Global infrastructure hub.

Goodland, R. (1995). "The concept of environmental sustainability." Annual Review of Ecology and Systematics 26: 1–24.

Goodland, R. and H. Daly (1996). "Environmental sustainability: Universal and non-negotiable." Ecological Applications 6(4): 1002–1017.

Gross, C. (2007). "Community perspectives of wind energy in Australia: The application of a justice and community fairness framework to increase social acceptance." Energy Policy 35(5): 2727–2736.

Gurara, D., V. Klyuev, N. Mwase and A. F. Presbitero (2018). "Trends and challenges in infrastructure investment in developing countries." International Development Policy| Revue internationale de politique de développement (10.1), Article 10.1. https://doi.org/10.4000/poldev.2802

Hall, J. W., A. Otto, A. J. Hickford, R. J. Nicholls and M. Tran (2016). A framework for analysing the long-term performance of interdependent infrastructure systems. The Future of National Infrastructure: A System-of-Systems Approach. A. J. Hickford, J. W. Hall, M. Tran and R. J. Nicholls. Cambridge, Cambridge University Press: 12–28.

Hart, H. L. A., J. Raz and L. Green (1961). The concept of law, Oxford University Press.

Hicks, J. R. (1975). Value and capital: An inquiry into some fundamental principles of economic theory, Oxford University Press.

IEA (2023). The world's top 1% of emitters produce over 1000 times more CO2 than the bottom 1%.

Infrastructure Australia (2021). Sustainability principles: Infrastructure Australia's approach to sustainability, Infrastructure Australia.

IPCC (2023). Summary for Policymakers. Climate Change 2023: Synthesis Report. I. a. I. t. t. S. A. R. o. t. I. P. o. C. C. C. W. T. Contribution of Working Groups I, H. Lee and J. Romero. Geneva, Switzerland, IPCC: 1–34.

Josa, I. and A. Aguado (2019). "Infrastructures and society: from a literature review to a conceptual framework." Journal of Cleaner Production **238**: 117741:1–117741:22.

Kuehn, R. R. (2000). "A taxonomy of environmental justice." Environmental Law Reporter News & Analysis **30**(9): 10681–10703.

Lange, G.-M., Q. Wodon and K. Carey (2018). The changing wealth of nations 2018: Building a sustainable future, World Bank Publications.

Macquarie, C. (1996). "The CCH Macquarie dictionary of law." Sydney, CCH Australia Limited.

Malthus, T. R. (1986). "An essay on the principle of population (1798)." The Works of Thomas Robert Malthus, London, Pickering & Chatto Publishers **1**: 1–139.

Manaster, K. A. (1995). Environmental protection and justice: readings and commentary on environmental law and practice. Cincinnati, OH, Anderson Pub. Co.

Mensah, J. and S. Ricart Casadevall (2019). "Sustainable development: Meaning, history, principles, pillars, and implications for human action: Literature review." Cogent Social Sciences: **5**(1). 10.1080/23311886.2019.1653531

Montaldo, C. R. B. (2013). "Sustainable development approaches for rural development and poverty alleviation & community capacity building for rural development and poverty alleviation." Wonju, Yonsei University.

Pepperdine, S. (2001). "Social indicators of rural community sustainability: An example from the woady yaloak catchment." The Future of Australia's Country Towns. Center for Sustainable Regional Communities, Australia's Online Library of Regional Research. La Trobe University, Bendigo, Victoria.

Pfeiffer, A., C. Hepburn, A. Vogt-Schilb and B. Caldecott (2018). "Committed emissions from existing and planned power plants and asset stranding required to meet the Paris Agreement." Environmental Research Letters **13**(5).

Porter, M. E. and C. Van der Linde (1995). "Toward a new conception of the environment-competitiveness relationship." Journal of Economic Perspectives **9**(4): 97–118.

Pouw, N. and J. Gupta (2017). "Inclusive development: a multi-disciplinary approach." Current Opinion in Environmental Sustainability **24**: 104–108.

Rawls, J. (1971). A theory of justice. Cambridge, MA and London, England, Harvard University Press.

Ross, D. (2013). Social sustainability. Encyclopedia of corporate social responsibility. S. O. Idowu, N. Capaldi, L. Zu and A. D. Gupta. Berlin, Heidelberg, Springer Berlin Heidelberg: 2245–2249.

Ruan, Z., C. Wang, P. M. Hui and Z. Liu (2015). "Integrated travel network model for studying epidemics: Interplay between journeys and epidemic." Scientific Reports **5**: 11401.

Ruttan, V. W. (1991). "Sustainable growth in agricultural production: Poetry, policy and science." Paper presented at the Seminar on Agricultural Sustainability, Growth, and Poverty Alleviation: Issues and Policies. Eds Vosti and Reardon, International Food Policy Research Institute, Feldafing, Germany, September 23–27, 1991.

Schlosberg, D. (2004). "Reconceiving environmental justice: Global movements and political theories." Environmental Politics **13**(3): 517–540.

Shahzalal, M. and A. Hassan (2019). "Communicating sustainability: Using community media to influence rural people's intention to adopt sustainable behaviour." Sustainability **11**(3): 812. https://doi.org/10.3390/su11030812

Singh, R., L. Mathiassen, M. E. Stachura and E. V. Astapova (2010). "Sustainable rural telehealth innovation: a public health case study." Health Services Research **45**(4): 985–1004.

Syme, G. J. and B. E. Nancarrow (2001). "Justice, sustainability, and integrated management: concluding thoughts." Social Justice Research **14**(4): 453–457.

Thacker, S., D. Adshead, M. Fay, S. Hallegatte, M. Harvey, H. Meller, N. O'Regan, J. Rozenberg, G. Watkins and J. W. Hall (2019). "Infrastructure for sustainable development." Nature Sustainability **2**(4): 324–331.

UN General Assembly (2015). Transforming our world: the 2030 Agenda for Sustainable Development.

UNU-IHDP. (2012). Inclusive wealth report 2012: measuring progress toward sustainability, Cambridge University Press.

6 Risks, vulnerability, and resilience in infrastructure

6.1 Introduction

Although infrastructures play a central role in every society due to the services they normally provide, infrastructures can be harmful to society as well as the environment. Infrastructures could not only be vulnerable to natural disasters but also could be unsustainable due to unbearable financial burdens (Thacker, Adshead et al. 2019). The harmful impacts of infrastructures on society and the environment may come from any phases over and beyond the project lifecycle. The lifecycle of a typical construction project starts with the development of the concept and goes over planning and budgeting, construction, operation and maintenance, disinvestment and recycling, and the consequential legacy of the project into society (Thacker, Adshead et al. 2019). For instance, power plants or cement production infrastructures could lead to a massive emission of CO_2 as a result of which the risk and vulnerability of the society to air pollution and climate change would be increased. Transportation infrastructures could facilitate access to natural resources, and this will lead to a more substantial depletion of precious non-renewable resources. Research revealed that on Australian roads alone, an estimated number of around 10 million animals are subjected to car accidents annually (Parrott 2020), as a result of which genetic resources are under strain. On the other hand, infrastructures could also promote pro-environmental actions by substitution of the so-called grey infrastructures with green ones (Economy and Climate 2016) or by treating the sewage through reed beds wetlands and ponds (Scholz and Lee 2005).

Another important issue with the infrastructures is that in some cases they can increase the vulnerability to natural or human-originated hazards and disasters. For instance, building an atomic power plant in a seismic active region, removing the vegetation close to infrastructures and providing a floodplain, or developing near flood-prone regions and mountainsides are examples that can pose vulnerability. Apart from that, our ever-increasing reliance on infrastructures represents a particular type of vulnerability to national security threats (Lewis 2019).

DOI: 10.1201/9781032622323-6

6.2   Types of risks in infrastructure projects

Infrastructure development requires intense capital expenditure and long-term investment. Therefore, assessing all the risks connected to an infrastructure project is crucial before making the final investment decision. Many investors are hesitant to engage in infrastructure projects because of these risks, but the loss rate may be decreased if the risk is recognised and acknowledged (Grimsey and Lewis 2002, Smith, Merna et al. 2014).

As depicted in Figure 6.1, risks associated with typical infrastructure projects can be divided into three groups: (i) Market-related risks brought on by the financial markets (Kumari and Kumar Sharma 2017); (ii) Institutional risks related to the regulations, restrictions, laws, partnerships, and public pressure; and (iii) Project related risks associated with the building and operation of infrastructure.

6.2.1   Market risks

The market risks are caused mainly by inadequate revenue hedging, rising financing costs, and unpredictable fluctuations in the financial market due

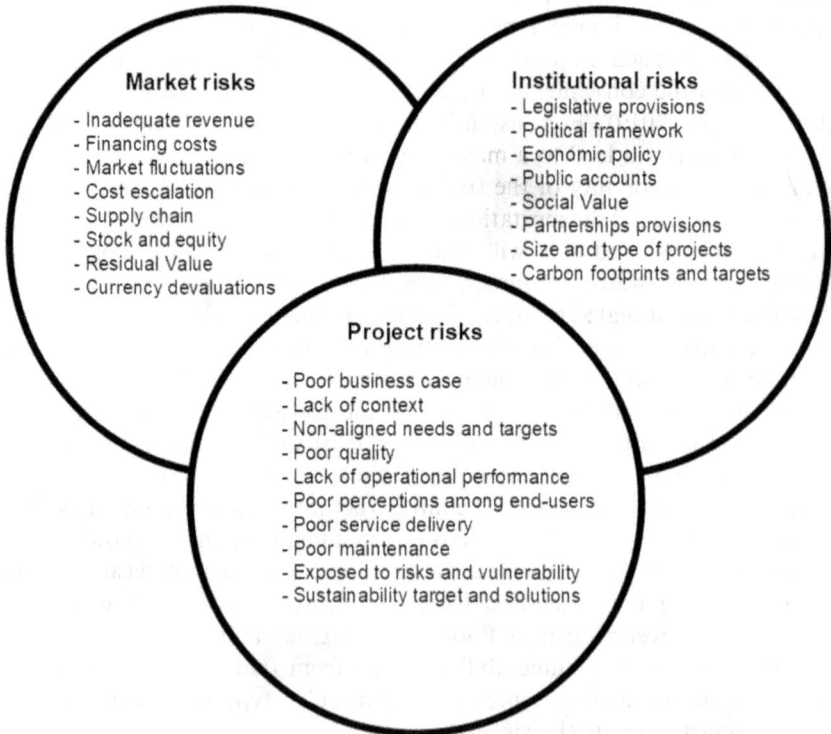

Figure 6.1 Risks in infrastructure projects.

to inflation and deflation effects. They are related to the financial elements of projects and could include some of the eminent risks such as risk of cost escalation, suffering losses due to the inability to recover capital including profitability, difficulty in meeting payments obligations or even erosion of financial values (known as liquidity risk). Some other associated risks may also include economic and accounting risks, stock declines or fluctuations (known as equity risks), force majeure or insolvency risks, credit and transaction risks, supply chain-related risks, lack of equity capital, residual value risks, and currency/devaluation risks (Nijkamp and Rienstra 1995, Grimsey and Lewis 2002, Merna and Njiru 2002).

6.2.2 Institutional risks

These risks result from the legislative and political framework of the project development environment and underlying economy, as well as from public sentiment and social value outcomes. Regulatory/political risks, currency inconvertibility and transfer hazards, contract breaches, riots and domestic disturbances are a few examples of this type of risk (Chapman 1997, Thobani 1999, Grimsey and Lewis 2002, Khan 2013). Infrastructure projects are capital intensive, for public projects, partnerships with the private parties are one of the preferred options in planning and development of public infrastructure. However, due to the varied interests of multiple parties and underlying institutional complexities of a particular country, numerous institutional risks may eventuate regarding the scope, size, and type of development projects.

6.2.3 Project risks

These risks are mostly connected to the development and maintenance of a project as a physical entity. These risks include the possibility that a project itself is not based on well-thought-through business cases and concepts that lack appropriate contextualisation. The physical facility resulting from the project may not offer the targeted services, programmes, or other interrelated deliverables, meeting the necessary level of quality and value (performance risk) for the end-users. In such a situation, there is a potential for suffering losses due to poor or ineffective procedures, rules, plans, or circumstances that interfere with corporate operations (operational risk), input risks, demand risks, and technical problems and operational risks (Grimsey and Lewis 2002, Merna and Njiru 2002, Matsukawa and Habeck 2007). It should be noted that operational risk could be caused by a variety of circumstances, including untrained employees, lack of knowledge, and education among the end-users, man-made or criminal activities like fraud, and compromise including natural disasters.

6.3 Risks in rural developments

While rapid urbanisation is a global phenomenon, rural regions are home to many impoverished families (Anderson 2003). Rural people are extremely susceptible because they frequently lack the tools and resources necessary to manage the risks arising from numerous activities, engagements, and operations. Therefore, one important tenet of an efficient and long-lasting rural poverty-reduction plan (SDG Goal 1 – No Poverty) is suitable risk-prevention strategies across all facets of rural life including provisions of risk-management solutions.

The context and associated risks affecting the rural communities are unique. Some of the common contexts of rural risks are shown in Figure 6.2. As seen, under the domestic context, some of the underlying risk events are due to family violence, abuse, and discrimination among different age groups, genders and sometimes also due to disability or mental illness, etc. Risk events in financial contexts are usually due to the lack of sufficient jobs and income-earning opportunities which eventually could trigger hatred and jealousy among some sections of the rural community. Being an agrarian community, reliance on nature is a common phenomenon and thus climate-related risks are quite prevailing in rural settings. Unprecedented fluctuations of air and soil temperature along with prolonged floods and drought due to climate change affect the

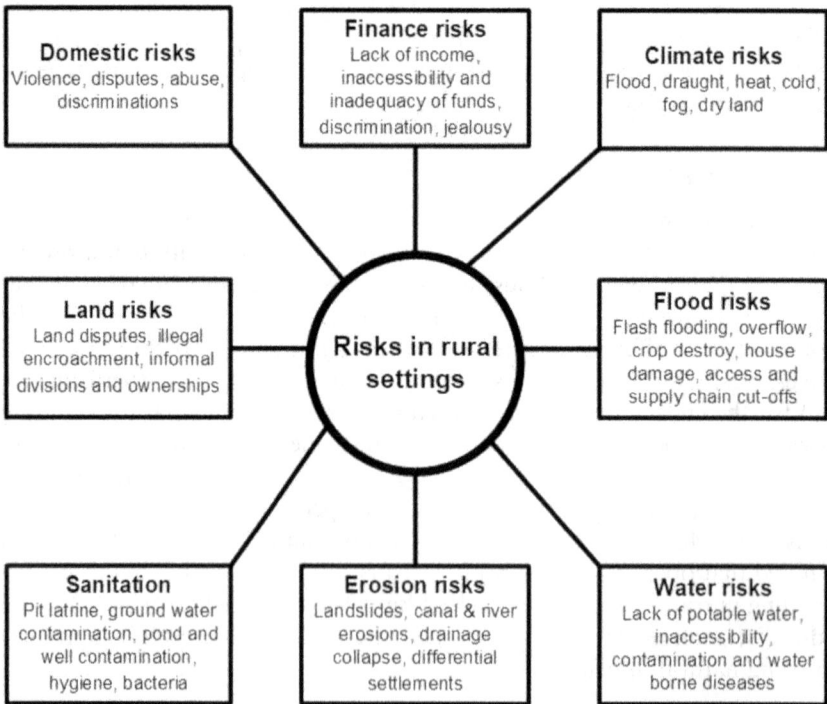

Figure 6.2 Common contexts of rural risks.

Image 6.1 Severe inundation of the natural park during Monsoon season in Assam.

livelihood of rural communities in a disproportional manner compared to their urban counterparts. Similarly, frequent flood events that cause flash flooding, and overflows trigger crops and property damage and cut off the communities from one another including blockages in the supply chains (Image 6.1).

Water-related risk is one of the most common risks the rural community face more than the other risks and this is due to the lack of infrastructure facilities for providing safe and clean drinking water to the population. Often the drinking water gets contaminated due to a lack of provisions for filtering and safe supply in the rural areas which then triggers waterborne diseases in the community. Erosion is one of the most significant risks in rural areas especially during Monsoon seasons when quick drainage of runoff water is not usually possible due to lack of drainage facilities or ill-maintenance of the waterways. Water clogging is one of the common causes of landslides and erosion as there is usually not any engineered or nature-based intervention to prevent the same. Sanitation risks arise from the pit latrines with the high possibility of groundwater, pond, and wells contamination which eventually raises the chances of hygienic conditions including breakout of the diseases. Land risks usually arise from land disputes among family members due to a lack of formal land registration and legal support. Land dispute also occurs in neighbourhoods in both agricultural farmlands and residential boundary fronts and these disputes are usually due to a lack of necessary systems including education and training among the common public.

Thus, context-specific rural risk management frameworks are particularly important to be in place for reducing vulnerability, especially for the people physically involved in high-risk activities. For a framework to be effective, it must include a variety of risk management techniques and execution plans (e.g. prevention, mitigation, coping) supported by active public participation (e.g. informal, context, or situational-based, public).

6.3.1   *Risks associated with rural development*

Any infrastructure-related development activity is a risky business within which many actors have different exposures to underlying risks commensurate with the potential roles to play (Anderson 2003). Besides, in the agrarian society where a large portion of the rural population resides, the intrinsic risks associated with agricultural activities are vividly obvious. The billions of farmers and other business owners who work in rural areas, as well as those who purport to represent them in the various spheres of influence, such as legislative or administrative domains, are subject to a plethora of unpredictable uncertainties. As a result, risk has widespread effects on rural development, and its impact at multiple levels of operations is enormous. The nature and intensity of the risk that farmers and other rural populations face vary substantially, depending on the agricultural system, local rural features, and climatological variations including infrastructural, regulatory, and institutional environments. The inherent complexity of these multiple dimensions entangling rural farmers makes risk management quite challenging for governments, NGOs, and international organisations. Besides, studies concerning regional economics give little attention to the impact of risk (Freshwater 2015). Regional economics literature either neglects to include risk in their topic index or just briefly mentions it in their discussion (Freshwater 2015). Similar to this, discussions of possibilities for local and rural economic development are sometimes constrained in terms of decisions that have particular intents or stipulated outcomes (Freshwater 2015). For example, communities or regions only can select from various tactics, such as industrial recruiting (inward investment), amenity upgrades, infrastructure investments, initiatives to encourage local entrepreneurship, or reinforcing already-existing company clusters. The emphasis is on choosing the "optimal" development approach, implicitly supposing that all methods give the same results or have equiprobable outcomes. However, the risk exists for every strategy and investment, which can vary significantly among alternative investments. In other words, when comparing development plans, discrepancies in the range of possible outcomes should be taken into account just as much as those in the projected outcome level (Freshwater 2015).

6.3.2   *Mitigating risks and underlying strategies*

Risk management needs to be a component of every rural development plan keeping the community's interests at the core, since the results of development initiatives are unpredictable, and may impact the communities with differential consequences. As a result, the theories of resilience and vulnerability are helpful when looking at different development options. When regions choose their investments, multiple criteria such as the rate of return and the intrinsic volatility of the returns need to be considered and weighed accordingly (Freshwater 2015). In several studies, Kostov and Lingard (2001, 2003) specifically address this issue by stating that risk reduction is one of the critical effects of rural development. Their main claim is that the high extent of economic specialism observed in rural areas with low levels of development leads to a significant amount of risk or vulnerability at the individual and group levels. When economic growth is necessary for rural development enabling local economy diversification, due consideration must be given to lowering risks and boosting resilience. The argument presented by Acemoglu and Zilibotti (1997) is quite similar in the context of macroeconomic growth and diversified development support.

When it comes the risk management, the scale, likelihood, and eminence are all important to consider. Some low-likelihood, high-consequence incidents may likewise be best ignored since they are both highly unlikely, and the expense of minimising exposure would be too expensive to outweigh the potential loss. For instance, every building intending to defend a region from a 100-year flood event may need significantly expensive design specifications. While such a 100-year powerful flood event, if occurs will overwhelm the control mechanism, due to less likelihood of occurrence and lesser eminence, a decision may be taken for a less expensive design specification while keeping the best interest of the community at large.

Some low-likelihood, high-consequence incidents may likewise be best ignored since they are both highly unlikely, and the expense of minimising exposure would be too expensive to outweigh the potential loss. Depending on the circumstances, at times it could also make sense to forgo the riskier options when a region offers several opportunities that all provide comparable results, but some are less hazardous. For instance, GM crops may allow farmers to increase yields while lowering production costs, but they do so with the risk of unfavourable effects on people's perceptions and to some extent the environment. Abandoning GM crops eliminates both the advantages and the risks. However, the opportunity cost for a complete elimination could be much higher than the risks being eventuated. When a specific risk is rejected, the option for that hazardous opportunity to become a part of the portfolio of activities is also forfeited. If the rejected risk choice has a negative correlation with the current portfolio, adding it would have decreased overall risk. The key takeaway from this situation is that portfolio risk, as opposed to individual hazards, needs to be the emphasis of risk mitigation efforts.

One of the most widely accepted approaches in managing high-consequential risks is through diversification (Freshwater 2015), whereby building a portfolio of risky behaviours reduces overall risk exposure. Diversification often causes the total return to decrease (Freshwater 2015). If projected returns differ among possibilities, specialising in the opportunity with the greatest expected return will yield the highest return over time (Freshwater 2015).

6.3.3 *Diversification of rural business activities*

Diversification of rural business activities is one of the important strategies in mitigating community and location-based risks. In a situation where a farming community growing a similar crop, there may be significant risk for individual farmers arising from a combined adversity of risk instances. One such risk instance could be that the farmers may face the same weather occurrences due to their proximity to one another, and since they grow the same or similar products, all the farmers will be exposed to similar market conditions and underlying fluctuations or uncertainties. Furthermore, the town may experience the same disturbances as the farmers because agriculture may be the pillar of its economy. In such a situation, the risk may be decreased for farmers and the region if the community diversifies its local economy by including an unrelated industrial sector and offers farm households off-farm work possibilities. The limited potential to mitigate risk via diversification is what distinguishes the situation of rural development from the situation of national growth (Freshwater 2015). While the rural economy is usually modest by definition, the lack of diversity of distinct company types is caused by the restrictive workforces, in terms of the number of employees and the accessible skills. This is particularly true if a business must operate at a minimal adequate size to be competitive. As a result, diversity is typically a poor strategy for achieving rural resilience. This necessitates the discovery of alternative risk-mitigation strategies.

The development of regions and communities is intrinsically a matter of chance due to the pervasiveness of risk. However, investments can change the likelihood of some possible outcomes or even the set of possible possibilities, which can change the odds of success (Freshwater 2015). A spectrum of individual economic possibilities is, in some ways, formed by the particular group of businesses in a town. Additionally, community economic performance is based on how effectively this portfolio performs. Although there is a particular distribution of future outcomes for any individual firm, the community's aggregate outcomes are determined by the particular portfolio of enterprises (Freshwater 2015). The connection between the risks impacting various enterprises is mostly reflected in the risk for this community portfolio.

This implies that a community must consider how the chances for new businesses or activities connected to the risk profile of their current economic structure rather than just evaluating the prospects of each particular operation. Suppose the risk distribution of a rural enterprise is negatively associated with the risk profiles of other enterprises. The high-risk enterprise may lower the overall community portfolio risk in that case. For instance, towns that depend on the forest have attempted to offer adventure tourism as a new pastime in several countries, including Scotland, Finland, Canada, and the USA. Although the advantages of tourism are pretty unpredictable, they are not significantly connected with money being made from timber logging (Freshwater 2015). Therefore, having two reasonably risky occupations might lead to less aggregated risk for the communities.

Important new insights can emerge when conventional economic development methods are viewed in the context of risk management (Freshwater 2015). Take an industrial cluster as an example. Advocates for clusters emphasise the advantages of coordinating small businesses' operations with close ties, either because they are all involved in the same industry or a local supply chain. The partnership offers significant chances for shared learning, marketing efforts, and exchanging employees and other resources. However, from a risk viewpoint, the close integration of enterprises might raise risk compared to a situation where firms are not integrated (Freshwater 2015). This is perceived as enhancing the yield, or advantage, to the firms and the region. A negative incident affecting one enterprise in the cluster now affects the others with a ripple as well. Although clusters may offer better revenue levels, they can also make businesses and the region more vulnerable.

Similar to this, extra risk ramifications are associated with the intention of people and organisations involved in economic development to emphasise multiplier effects as a means of capturing more local advantages from adding a new enterprise (Freshwater 2015). For instance, an agricultural town that expands a fertiliser company or a biofuel factory will profit from the local supply chain. However, if the harvest fails, all the businesses will suffer. The susceptibility grows with the degree of company interdependence, which raises income and employment multipliers (Freshwater 2015). Any local supply chain component that fails affects all the chain's participants, not just the failing business. The likelihood of a greater predicted growth rate in the local economy is essentially increased with strong local supply chains, but only at the expense of a higher level of risk (Freshwater 2015). This is consistent with a risk-return trade-off being present.

Finally, it is generally recognised that few new entrepreneurial operations ever employ more than one or two individuals and that new entrepreneurial endeavours have significant failure rates in their initial years (Headd 2003). Because it is challenging to identify potential business owners and motivate them to act on their ideas, and since the rates of return to the region or community are modest given the effort required, this has tended to wane the interest of those in charge of local economic development in entrepreneur-based strategies. However, suppose the risk is primarily unconventional at the level of each business owner, meaning that the success or failure of any one firm is unrelated to the fate of others from the perspective of the community or region. In that case, the aggregate risk from this strategy is lower than the risk faced by any entrepreneur. Each company's chance of surviving is a binary event; either it fails or succeeds. However, in a community or region with numerous business owners, the majority of businesses will probably survive, and the area will be better off as a result.

The consequences of systemic risk and idiosyncratic risk must be distinguished from the standpoint of local economic growth (Freshwater 2015). As stated earlier, economy-based diversification is frequently pro-moted as a strategy for growth (Freshwater 2015). Diversification, however, can only deal with idiosyncratic risks or risks unique to a particular company. On the other hand, business cycle risk is systemic and unverifiable since it impacts all enterprises (Freshwater 2015). Similar to this, businesses in the community or area are connected through supply chains if they exchange a sizeable portion of their total output with one another. In good times, these supply networks' greater local multipliers are beneficial, but in bad ones, when unfavourable impacts spread across the supply chain, they are undesirable. The primary source of risk that businesses in the community or area face needs to be unique for diversification to be successful, ensuring that one business' success or failure has little effect on the others.

Understanding the collective risk instances among the business partners or enterprises is one of the key requirements for assessing the eventual risks and consequences of a particular business operation. Figure 6.3 highlights a hypothetical network-based risk association among the business partners (so-called stakeholders) involved in a typical business-related process. While the size of nodes represents the degree of importance of the stakeholders in the relative term, the directional arrows and thickness depict the flow and volume of information flowing from one stakeholder to the others. This visual depiction of the network connections between the stakeholders provides clear dependency relations in transactional business processes. The risks associated with each stakeholder in the network can easily be assessed from such relational dependencies which eventually assist in minimising or mitigating vulnerability appropriately (Doloi 2018). An intrinsic characteristic of recognising the underlying risks in any operations and appropriate responses to reducing the vulnerability is crucial in development solutions in rural settings (Freshwater 2015).

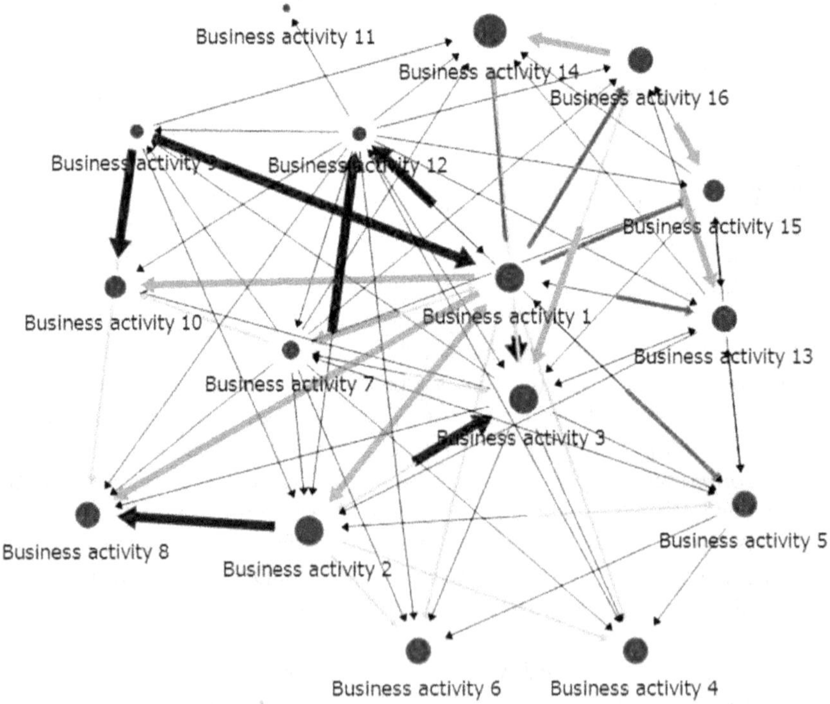

Figure 6.3 Responsibility sharing network among stakeholders in a typical business-related process (by frequency of transaction or communication).

Economics acknowledges that specialisation may boost output and raise expected profit. However, specialisation also entails more significant risk than an economy with a broader range of industries. This is consistent with a risk-return trade-off being present (Freshwater 2015). There can be a threshold to how much variety is tolerable in an area or community unless risk minimisation is the primary objective (Freshwater 2015).

6.4 Vulnerability concept and its drivers

6.4.1 *Concept of vulnerability*

Depending on the disciplines and/or organisational settings, the idea of vulnerability could vary across project types and operational environments. According to a general definition, vulnerability refers to how exposed people are at risk, their capability to respond to those threats, and the repercussions in terms of a loss of well-being. The degree of vulnerability is greatly influenced by a person's capacity to deal with challenging external circumstances and the social, economic, political, and environmental systems wherein they reside. In this regard, there are also various definitions presented in the literature for the concept of vulnerability. For instance, the World Bank defines vulnerability as the probability or risk of being hit by poverty now or falling into deeper poverty in the future. "It is a key dimension of welfare since a risk of large changes in income may constrain households to lower investments in productive assets – when households need to hold some reserves in liquid assets – and in human capital" (Lazarte 2017). International Federation of Red Cross and Red Crescent Societies (IFRC) (2012) defines it as "the diminished capacity of an individual or group to anticipate, cope with, resist and recover from the impact of a natural or man-made hazard. The concept is relative and dynamic. Vulnerability is most often associated with poverty, but it can also arise when people are isolated, insecure, and defenceless in the face of risk, shock or stress". United Nations Environment Programme (UNEP 2002) wrote "Vulnerability is the manifestation of social, economic and political structures and environmental setting. Vulnerability can be seen as made up of two elements: exposure to hazards and coping capability. People having more capability to cope with extreme events are naturally also less vulnerable to risk". Blaikie et al. (2003) argue that there are distinct approaches to defining vulnerability. For instance, one can utilise the demographic approach, in which vulnerability is a characteristic of a group, including children, women, the elderly, the impoverished, individuals with disabilities, and others, or the status of particular socioeconomic categories of people. The other way to do so is the taxonomy approach, wherein classifications are based upon perceived causative agents, including physical, social, economic, and environmental factors. Investigating the existence of vulnerability in everyday life, and looking at the realities, temporal aspects of risk, and the context is another way to do so. Some researchers also believe that the other way is that the members of the community define their weaknesses and strengths and contextually assess the relative weak points (McGee and Penning-Rowsell 2022).

6.4.2 Introduction to concepts of disadvantaged and rural populations

In this context, "disadvantage" refers to specific circumstances that limit a population's ability to take advantage of opportunities to enhance their living standards, access to decent jobs, and working conditions because of geographical, governmental, cultural, socioeconomic, and/or environmental factors. Social and economic inequities in rights and privileges are a significant cause of discrimination towards vulnerable communities and can restrict the use of resources (Lazarte 2017). Access to resources, markets, and public services like healthcare is further hampered by inadequate rural infrastructure and services for transportation, power, and clean water, extending the time required for household and care work (Lazarte 2017). Indigenous and tribal communities will have less access to financial possibilities, especially those geared at lowering their vulnerability to specific environmental risks, due to the lack of clarity in words that define property borders and ownership rights (Lazarte 2017). Social conventions and attitudes that prevent women from accessing the same relationships may cause their networks to be poor, even though these networks are crucial for connecting them to job prospects (Lazarte 2017). Youth in rural areas have the potential to be real change agents. Their potential is, however, under-valued and ignored in domestic and global development programmes (Lazarte 2017). When youth employment difficulties are discussed, the emphasis is frequently on more educated urban kids, leading to low-quality instruction, a lack of training curricula relevant to local requirements, and "relatively" high tuition costs (Lazarte 2017). Because there are few prospects for educated individuals to use their abilities constructively in rural regions, emigration is encouraged, creating a "rural youth drain" (Lazarte 2017). Many migrant labourers are forced to work in hazardous conditions in agriculture, one of the riskiest industries (Lazarte 2017).

Socio-economic empowerment is, therefore, the process of ensuring equality and possibilities for disadvantaged segments of the population, either by themselves or with the aid of others who share their access to these chances, to overcome disadvantageous situations. Empowerment entails visibility, a voice, improved understanding of rights and duties, and frequently, it entails vigorous advocacy efforts.

On the other hand, marginalisation describes a scenario where a group of individuals have been pushed to the periphery of possibilities and advance-ment due to various factors, including but not limited to ethnicity, citizen-ship, age, gender, origin, education, culture, and/or economic position (Lazarte 2017). Geographically isolated people are less economically compet-itive due to high transaction costs, a limitation of services, an absence of access to other marketplaces, and a lack of awareness of available alternatives

(Lazarte 2017). In addition, isolation and distance are frequently linked to weak legal systems and inefficient labour market protections (Lazarte 2017).

Social and economic inclusion is the best method for enhancing the conditions under which individuals participate in society, safeguarding their rights and giving them chances to better their situation (Lazarte 2017). Promoting rights and entitlements literacy, encouraging diversification to combat economic dependency, updating or changing policies and improving mechanisms for their implementation, public awareness campaigns, and enabling access to services, knowledge, and financing are a few examples of socioeconomic inclusion measures. The need to promote networking agreements, provide novel techniques in underdeveloped institutions, and lessen isolation through rural accessibility initiatives must be firmly focused (Lazarte 2017).

6.4.3 Addressing vulnerability through risk management

Some demographic groups are more prone than others to be susceptible to hazards, and they may also lack the resilience strategies needed to foresee, manage, resist, and recover from risks' effects. These vulnerable populations must be the focus of any government effort or policy action. Building resilience entails creating more robust "risk management" tactics, which can be official or unofficial and include (Lazarte 2017) adaptation, reduction, sharing, avoidance, retention, and preparedness.

One of the key requirements for understating risks and addressing vulnerability in a particular community is comprehensive demographic data and underlying analytics, depicting the risk instances in both individual and dependency settings. Depending on the demographic distribution coupled with socio-economic status, risks and vulnerability may differ. Thus context-specific intervention planning and appropriate management strategies are important to devise effective risk management and addressing vulnerability, ensuring the growth and well-being of the community. The following section discusses further the vulnerability and the underlying vulnerability dimensions from the rural communities' perspectives.

6.5 Vulnerability in rural communities

Rural people are defenceless due to low agricultural harvests, water shortages brought on by drought and other social problems, including caste operationalisation. With its detrimental effects on rural output and shifting parts of subsistence economies, climate change negatively influences rural populations. This lessens the incentives and capacity for populations to stay in rural regions.

The majority of the world's poorest people reside in rural regions and rely heavily on farming and other climate-sensitive industries for their existence. Their lives, properties, and incomes remain very exposed to climatic whims due to limited access to resources for coping with and adapting to climate-related shocks. This has important ramifications for the adaptation and mitigation tactics used to lessen susceptibility. Some of these factors and potential impacts are discussed below.

6.5.1   *Climatic (natural) factors affecting vulnerability in rural and regional areas*

Agriculture is innately susceptible to climatic fluctuation and change since it serves as the primary source of income for rural communities in developing nations. In many arid and semi-arid parts of the developing world, drought is commonly cited as a severe danger to agricultural systems and the livelihood security of farm people. The prospect of climate change makes the adverse effects of drought even worse. Increasing farm households' adaptive ability is essential to reducing the detrimental effects of the drought on their way of life. Farm families are finally compelled to give up farming if rural livelihoods become unstable. A good case study consisting of 274 farm families has investigated the livelihood vulnerability of rural communities of Iran to drought (Keshavarz, Maleksaeidi et al. 2017).

Flooding is one of the other main outcomes of climate and is viewed as a climatic factor affecting the vulnerability of the rural population. However, until now, research on the dynamics of vulnerability has either been regionally or nationally oriented or evaluated the impact of the city's size on the city's vulnerability. More consideration needs to be given to how a city's size affects nearby rural regions' vulnerability. In this regard, Jamshed and his co-workers (2020) considered this subject in rural Pakistan investigating 325 flood-affected rural communities in the vicinity of three cities in Punjab, Pakistan.

The assessment of the intrinsic vulnerability of rural populations in the Kimsar region to environmental risks is the main objective of the study of Rajesh et al. (2014). To measure the communities' intrinsic vulnerability to future harm, they created a new conceptual framework and found risk-generic socio-economic indicators. To gather data on the designated indicators, they surveyed all the households in the chosen villages. Eight sub-components of inherent vulnerability were represented by these variables, including gender, marginalised populations, economic capability, dependency on environmental resources, and lack of access to water, housing, information, and connection (Rajesh, Jain et al. 2014).

For public health and socioeconomic development, clean water, sanitation, and hygiene (WASH), including drainage services, is crucial. However, access is still insufficient and unequal in low- to middle-income nations like South Africa (Abrams, Carden et al. 2021). Rural and small towns in South Africa typically rely on a constrained and climate-sensitive economic basis (such as farming), have a small population and are situated in regions where transportation issues might raise the risks associated with WASH access (Abrams, Carden et al. 2021). In addition, hydrological cycles are altered by climate change, which can decrease WASH access and make already vulnerable areas more susceptible to the interconnected effects of droughts and flooding (Abrams, Carden et al. 2021).

In this regard, Abrams et al. (2021) used two case studies to examine climate risk and vulnerability assessment (CRVA) in a rural village in the northern Limpopo province and a small town in the Western Cape province. They embraced a multidisciplinary and interdisciplinary approach to investigate the requirements, impediments, and vulnerabilities concerning WASH in rural areas and small towns in South Africa (Abrams, Carden et al. 2021). Their comprehensive approach took into account both environmental and climatic conditions as well as socio-economic (economic, social, political, and governance) aspects to understand how these interrelate to obstruct access to WASH (Abrams, Carden et al. 2021). Frequent and severe droughts or floods caused by extreme weather conditions worsen groundwater supply, harm water infrastructure, and endanger lives reliant on agriculture. A lack of dependable transportation infrastructure increases flooding dangers since damaged roads to essential supplies are more likely to occur. In this regard, Abrams et al. (2021) concluded that inequitable access to WASH services is caused by high inequality, growing unemployment, and the Apartheid legacy of a segregated service delivery system. The interconnected manner in which South Africa's historical, social, economic, governance and policy components are changing makes rural and small towns more vulnerable to WASH issues (Abrams, Carden et al. 2021).

There are also other studies in which attention has been paid to the climatic (natural) factors affecting vulnerability in rural and regional areas. For instance, Fahad and Wang (2020) have reviewed the reciprocal relationship between rural vulnerability and climate change in the context of Pakistan. Rahman and Hickey (2020) proposed a five-step framework to assess the context-specific vulnerability of rural livelihood. Khalid and her co-workers (2021) investigated the Hindu Kush Himalayan region in Northern Pakistan and proposed a holistic multidimensional vulnerability assessment framework considering socioeconomic, physical, institutional, physical, gender-related, and attitudinal factors.

6.5.2 *Economic (human) factors affecting vulnerability in rural and regional areas*

Several human-related/economic factors can affect the vulnerability of the rural and regional areas. For instance, the study conducted by Imai et al. (2015) investigated whether rural non-farm income in Vietnam and India impacts alleviating poverty and vulnerability. They have employed a treatment-effects model to account for the bias associated with sample selection (Imai, Gaiha et al. 2015). In both Vietnam and India, it was discovered that access to rural non-farm income considerably raised log per capita consumption or log mean per capita expenditure, which is consistent with its role in alleviating poverty (Imai, Gaiha et al. 2015). However, the overall effect was more significant in Vietnam than in India (Imai, Gaiha et al. 2015). They also added that in both nations, access to rural non-farm work also considerably lowers vulnerability, suggesting that shifting family activities into the non-farm sector might lower such risks. When they broke down non-farm sector employment by type, they discovered that the impacts on lowering poverty and vulnerability in both nations were significantly more important for sales, professionals, and clerks than for unskilled or physical labour (Imai, Gaiha et al. 2015). However, given that the rural poor do not have easy access to skilled non-farm work and that even manual or unskilled non-farm employment greatly decreases poverty and vulnerability in India and poverty in some years in Vietnam, this has important policy relevance (Imai, Gaiha et al. 2015).

6.5.3   *Effect of rural infrastructures on the vulnerability of communities*

Rural households' long-term reliance on agriculture as a source of income exposes them to economic vulnerability, contributing to poverty in rural areas. An extensive increase in accessibility to rural infrastructures would be an excellent place to start when trying to mitigate this risk. Moreover, given the significant reduction in transportation costs, rural populations would no longer be as cut off from essential social services. To entice private enterprises to launch activities in rural regions, a lobbying campaign and incentive scheme would be required. Expanding private businesses will protect rural families from exposure to vulnerability and act as a driver for the growth of microenterprises. As long as these businesses had a robust social responsibility policy to prevent inequality, sustainable rural development would ensue (Barrios 2008). Ecological integrity would also require a plan for managing natural resources.

Community engagement is essential in selecting development initiatives since it may reduce resource waste on unproductive projects and allow for resource allocation to more productive applications. Rural road construction should be effectively packaged with services, support, and capacity-building initiatives (Barrios 2008). This may increase the need for more infrastructure and services, leading to a dynamic evolution of crucial components for rural development. In addition, bundles of interventions increase rural families' production efficiency at all production phases, both on and off the farm. Interventions in rural development should concentrate on reaching out to the community's more vulnerable members, notably the farmers. Interventions should progressively free people from reliance on agriculture without jeopardising their ability to get food.

To provide a sustainable approach to rural development, public infrastructure investments and consumption taxes can complement one another by continuously providing new infrastructure and maintaining old infrastructure (Barrios 2008). In addition, the socialised users' fee system can stop the income gap in rural regions from worsening. However, choosing a proper and adequate foundation for the socialised users' price rate is crucial. A questionable rate decision could be interpreted as a barrier to access or as a sign that the government is not sincere about supporting rural development by a section of the rural community. In the long run, this may exacerbate societal problems rather than reduce inequality.

Barrios (2008) classified the development interventions into four categories including infrastructures and institutions for production support as well as credit allocation, physical infrastructure such as roads, capacity building such as training and dissemination of information, and support services such as social services, and marketing.

Isolation, a shortage of or insufficient provision of necessities, subpar social and health services, etc., are characteristics of rural places (Barrios 2008). To allow the other difficulties to be overcome, isolation must first be addressed. Access to essential supply sources and market locations is facilitated by farm roads. Roads are anticipated to make it easier to lower the cost of moving agricultural inputs and getting the output to market (Barrios 2008). Glaeser and Kohlhase (2004) reported that a good road network could reduce the cost of transporting goods by up to 90%. Infrastructure is essential from an economic standpoint, but there may also be some unfavourable societal effects. For instance, dams are thought to aid in the sustainability of irrigation. However, on the other hand, as Duflo and Pande (2007) highlighted, they are expensive and contentious, and little is understood regarding their broader implications.

Investment in the development of basic infrastructure is necessary to move towards social development (Hemson, Meyer et al. 2004). Such infrastructure can support rural development and, therefore, can reduce poverty. In addition, rural empowerment, which must promote civil society and public engagement in decision-making in a democratic culture, is directly related to rural development. The International Labour Organization examined the relationship between poverty and accessibility (ILO 2005). Poor, isolated communities have less access to necessities like health care and education, which are major risk factors that cause deprivation and lead to poverty. Infrastructure in rural areas is thought to make it easier to get these items. However, recent events have shown that, although infrastructure supply is vital, it is insufficient to reduce poverty (Barrios 2008).

It is necessary to recognise community engagement's contribution to improving local public service delivery. Development can either be facilitated or hampered by the interactions between the community, local authorities, and higher levels of government (Das Gupta, Grandvoinnet et al. 2003). The importance of community engagement stems from several factors, including the residents' knowledge and comprehension of the environment, the information gaps across households, and the fact that the results will directly impact them. Interventions that might lessen power inequalities among community members, such as land reform, the creation of nonfarm sources of income, etc., should be used to strengthen a development-directed state-community synergy. The vested interests of local administrators can be avoided by policies at the higher levels of administration, making them more responsive to family demands. Numerous elements, such as the creation of community demand (participatory) for improved public goods and services to promote development, can help institutional changes at the local and community levels (Barrios 2008). For example, this might involve empowering techniques like infrastructure development and capacity building in rural areas, which would reduce the cost of transportation and increase access to markets and farm supplies. In addition, the knowledge imbalance between suppliers (of inputs), traders (of produce), retailers (of food goods), and producers will be reduced, if not eliminated, through improved accessibility (Barrios 2008). Therefore, rural infrastructures decrease the vulnerability of the rural community as they play important roles in poverty reduction, demarginalisation, income diversification, promotion of entrepreneurship, and increasing the efficiency of production (Images 6.2 and 6.3).

Image 6.2 Severe flood water uproots trees in natural park during Monsoon season.

Image 6.3 Large-scale brick and mortar development in rural areas – a contrasting view to the village landscape.

6.6 Factors affecting resilience in communities

Resilience in the communities refers to the withstanding or coping power of the people against unknown extreme events that otherwise adversely affect the community's way of life. The idea of community resilience is becoming increasingly important in modern society, especially since it seems that many rural communities throughout the globe are steadily losing resilience in the face of problems like climate change, youth outmigration, and socio-economic disruptions (Rigg 2006, Kelly, Ferrara et al. 2015, Wilson, Schermer et al. 2018). The capacity of a system, group, or society to withstand, digest, adapt, and recover from shocks and disruptions, especially via the preservation and reconstruction of its core basic structures and functions, is known as resilience (Wilson 2012). Disaster magnitude is frequently determined by a combination of factors, such as economic hardship, destruction of the environment, the prolonged effects of prior events such as war, or a cascade of events brought on by systemic interdependencies that have hampered a community's capacity to react (Brown 2014, Wilson, Hu et al. 2018).

Hitherto most of the studies have focused on community resilience in developed countries (Rigg, Salamanca et al. 2012). However, about the fast-paced evolution of rural communities especially in developing countries, research on this subject would be necessary. To analyse the resilience of the communities, the underlying factors and context-specific impacts are important to understand. Figure 6.4 highlights some of the key factors affecting community resilience in rural settings (Image 6.4).

6.6.1 *Economic factors*

Economic factors affecting community resilience include income sources, property damage, loss and compensations, inflations and cost escalations, investment and opportunities, insurance and security, and diversification and accessibility. Resilience on the economic front is about inherent responses and the ability to adapt to these factors by the members of the community without being exposed to risks and vulnerability. For instance, the continuity of household income for livelihood support is one of the key indicators of how a particular member of a family is resilient against unforeseen events. Property damage refers to the minimisation of the severity of damage and regaining the functionality of the properties such as housing, educational institutions, health care centres, community and recreational centres, etc. Losses in monetary terms need to be minimal for increased resiliency and there should be adequate provisions for recouping the losses as compensations being paid to the affected parties. Inflation is linked to the rise in costs and fluctuations in prices while reinstating the damages from risk eventualities. In the event of any damage occurring from an unforeseen event, the post-event phase should present necessary opportunities for new investments and jobs for the community to engage. Third-party risk transfer through insurance schemes is quite a prevailing practice for enhancing security and

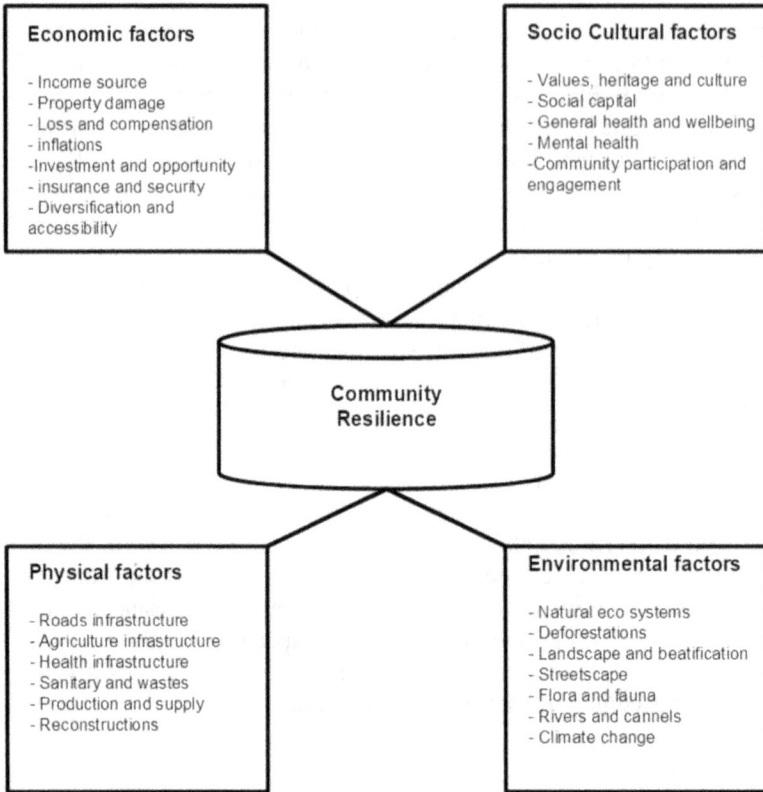

Economic factors

- Income source
- Property damage
- Loss and compensation
- inflations
-Investment and opportunity
- insurance and security
- Diversification and accessibility

Socio Cultural factors

- Values, heritage and culture
- Social capital
- General health and wellbeing
- Mental health
-Community participation and engagement

Community Resilience

Physical factors

- Roads infrastructure
- Agriculture infrastructure
- Health infrastructure
- Sanitary and wastes
- Production and supply
- Reconstructions

Environmental factors

- Natural eco systems
- Deforestations
- Landscape and beatification
- Streetscape
- Flora and fauna
- Rivers and cannels
- Climate change

Figure 6.4 Factors affecting resilience in rural settings.

Image 6.4 A tertiary institution building with a low-cost structure with a tin roof in a village in Assam.

resilience within the community. Regarding the new opportunities and broad accessibility, diversification strategies should be in place so that the community gets the choice to continually engage in professional roles even if there are risks involved in any particular operation or business.

A lot of studies have focused on the economic factors influencing the resilience of the community (Buikstra, Ross et al. 2010). However, there are still major debates on how the embeddedness of communities into economic networks, or the integration of the rural and regional economies into the more market-driven economies could affect the resilience of the community (Wilson, Hu et al. 2018). In other words, there is still an ongoing discussion on whether globalisation may support more resilient approaches by presenting a more comprehensive range of development prospects (Wilson 2012). Poverty (lack of resources to incorporate intrinsic resilient mechanisms), excessive reliance on outside sources of livelihood, or high-income inequality are economic factors that can aggravate the vulnerability of populations and undermine community cohesion, trust, and networks (Kelly, Ferrara et al. 2015). A good example of this subject can be found in the study of Wilson and his co-workers (2018) on the community resilience of rural China.

6.6.2 *Socio-cultural factors*

The social sphere is also essential for resilience because it controls the connections between a community's economic, cultural, and environmental elements (Wilson, Hu et al. 2018), according to regionally particular social connections and values. Relationships, trust, participation of young and old people, dispute resolution techniques, networking capacity, learning and communication routes, bonding and bridging capitals, collaboration, and community "cohesiveness" are key social elements affecting the resilience of the community (Rigg, Salamanca et al. 2012, Wilson, Hu et al. 2018). The reason behind this impact is that well-developed social processes are typically an indication of a high capacity to adjust to risks and disasters. Conversely, communities with poor social capital permit behaviours that increase vulnerability, and collective concerns are more likely to be disregarded by influential parties, leading to the fragmentation of community interests (Wilson, Quaranta et al. 2015). Young people leaving their communities is frequently a significant contributor to a poor social fabric and is a phenomenon that is directly related to economic concerns. In any social context, outmigration is a crucial aspect to consider since it is both a cause of vulnerability, such as the loss of social memory and skills and a reaction to deteriorating resilience. Outmigration leads communities to age and become gentrified, which impacts social dynamics, interferes with the generation gap, and disturbs social memory due to loss of experience and knowledge. Social memory can no longer be utilised to respond to shocks or disruptions after it has been gone (Wilson 2015). These social actors can negatively impact the resilience of the community while remittances could have some positive impacts (Wilson, Hu et al. 2018). Local institutions may not adapt due to migration, which can diminish bonding capital by decreasing community cohesiveness, reducing human interaction, and dissolving groups based on shared ideology and sets of laws.

Similarly, values, culture, heritage, rituals, beliefs, social norms, traditions, and customs are all included in the socio-cultural realm. Socio-cultural elements are significant because they comprise complex human behaviours passed down via instruction, imitation, and conformity rather than through genetic transmission or external pressure (Wilson 2012). Therefore, teaching and learning, as well as the moral aspect of community rules, customs, and rites that support the "how things are done" and penalise transgressions, are crucial to understanding the impact of cultural processes on community resilience. In the end, cultural processes enable people to do things inside a community they would not be able to do on their own. Moreover, ideologies are strongly related to how societal trends and preferences develop and how those changes impact how groups make decisions (Rigg, Salamanca et al. 2012). These, in turn, impact a community's social, economic, and ecological realms. Therefore, cultural influences may be considered the network of ideas that permeates societies and forms collective social awareness.

6.6.3 *Physical factors*

Physical factors are usually about the physical or built-up infrastructure facilities such as roads, bridges, tunnel infrastructure, agriculture-based infrastructure, health-related infrastructure, sanitary and wastes infrastructure, production and supply supported infrastructure including retrofitting, refurbishment and reconstructions. While these infrastructures act as backbones for running the community's normal life, in the circumstance where major risk events may occur, prolonged break-down of these infrastructures could have a detrimental effect on society. Thus to build resiliency in the community in this physical context, operational and functional risks are required to minimise so that the downtime is avoided. As the physical infrastructure systems are usually complex and there is a high degree of operational interdependencies between various types of infrastructures, proactive management of risks associated with the system is highly crucial. Probabilities and potential consequences of any failure events that may potentially bring significant disturbances to the operations of the infrastructure network need to be assessed carefully so that appropriate measures can be taken to ensure the required level of resilience in the community. Usually, the extent of disturbances and ripples exerted by the physical factors are quite widespread and thus strategies for building resilience encompass significant considerations above and beyond the local settings and operations.

6.6.4 *Environmental factors*

Environmental factors comprise the phenomena arising from natural ecosystems, deforestations, landscape and beatification, streetscape, flora and fauna, rivers, and canals including climate change. The risk events from these factors can impact community resilience (Wilson, Hu et al. 2018, Farahani and Jahansoozi 2022). Disturbances in the natural ecosystems can destroy the integrity of the natural environment around the community which could inflict psychological and mental irritations on some people, especially nature lovers including adolescents and elderly people. Similarly, deforestation, degradation of landscape and natural beauty including streetscape, local flora and fauna due to man-made disasters or natural calamities could exert significant risks and uncertain conditions on the community. Building resilience against such factors requires careful assessments of the environmental impacts including underlying strategies for minimising the disturbances and managing the same. Increasing extreme events due to climate change such as intense rainfall, and prolonged draughts affect the waterways, rivers, and canals including farmlands and agricultural produce. While the risks and vulnerability resulting from these factors are quite real and can affect almost every section of the rural community, building resilience against such events requires significant financial and physical resources at both community and governance levels.

6.7 Resilience in the infrastructure system

Disasters, whether man-made or due to natural calamities, affect people across urban and rural regions worldwide (Mao and Li 2018). According to the International Strategy for Disaster Reduction study (UNISDR 2015), natural catastrophes, including seismic events, flooding, blizzards, and cyclones, have resulted in yearly economic losses of more than USD 300 billion. Moreover, these natural catastrophes significantly cause infrastructure assets' vulnerability (Shi, Wen et al. 2019). As an illustration, the severe cyclone "Sandy" in New York in 2012 severely damaged the region's infrastructure and caused a significant economic loss to the city's citizens of over USD 70 billion (USDOC).

Resilience is typically characterised as a system's, society's, or region's capacity to quickly and effectively respond to, digest, adapt to, and rebound from disruptive occurrences (Engler, Göge et al. 2018). Infrastructure resilience (IR) is the capacity to recover entirely from catastrophes when infrastructure systems sustain localised damage due to highly disruptive natural or artificial occurrences (Ouyang, Liu et al. 2019). Due to the virtue of infrastructure systems being highly interconnected, the degree of dependency may determine how resilient they are to frequent disturbances and catastrophic catastrophes (Rahimi-Golkhandan, Aslani et al. 2022). According to Shakou et al. (2019), a city or regional area will be severely susceptible to disturbances and turbulent operations if its infrastructure systems are not resilient. Thus, it is crucial to maintain and improve the resilience of infrastructure systems, especially when individuals experience severe interruptions.

Many different models are being applied in practice for assessing resilience. However, the scope of most of these models is confined to natural disaster management. For instance, focusing on ecological aspects in disaster-prone areas, the Tobin Model (1999) assesses the community's resilience across three key areas, *risk reduction pattern, recovery pattern,* and *structural and demographic pattern.* Davis et. al. presented a model that uses the timeline of a community to improve resilience based on three distinct phases, *absorption and tolerance of pre-disaster stress and underlying dangers, involvement in wider community interaction,* and *changes in communities to make safe and resilient* (Davies, Streeter et al. 2018). The Baseline Resilience Index Conditions (BRIC) model can provide a comparative overview of dimensions in resilience baseline indices based on a context-specific disaster event (Camacho, Bower et al. September 2023). Encompassing the social, economic, institutional, and physical interventions for improving societies, the model provides a methodology and a set of indicators for mapping both existing and improved conditions and ensuring resilience accordingly. Drawing from the Author's own experience, Figure 6.5 shows the key processes for achieving IR which are briefly discussed below.

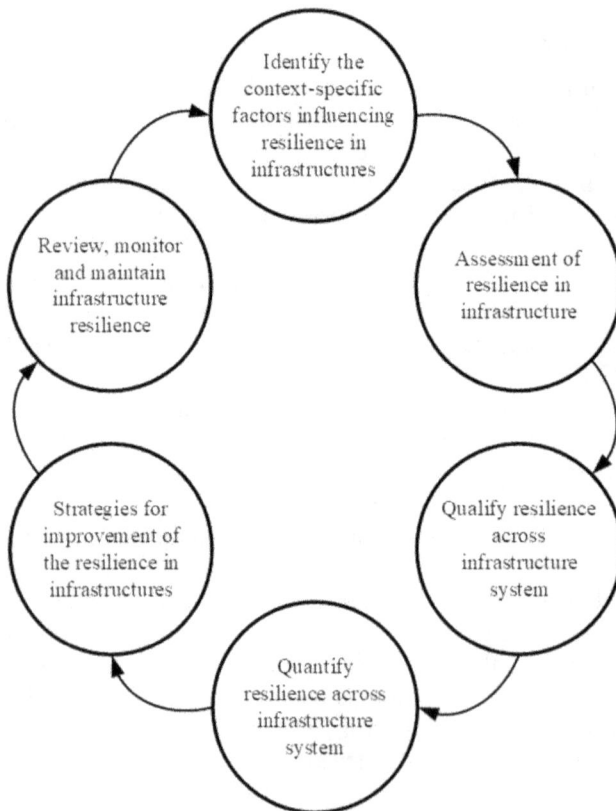

Figure 6.5 Achieving resilience in infrastructure systems.

6.7.1 Context-specific factors influencing resilience in infrastructures

The contexts and functions of infrastructure systems vary from one region to the others. Irrespective of the contexts, infrastructure systems are continually impacted by various circumstances and an accurate understanding of the difficulties affecting the IR is a challenging task (Faturechi and Miller-Hooks 2015). As a first step in building resilience, identification of the factors associated with a range of different contexts is highly crucial (Rehak, Senovsky et al. 2018, Splichalova, Patrman et al. 2020, Liu, Love et al. 2021). For instance, Bundhoo et al. (2018) employed resilience theory to recognise the problems of energy IR involved in repairing, reconstructing, and retrieving energy security of Small Island Developing States (SIDS) to catastrophic weather events, such as cyclones, hurricanes, and floods. They observed the robustness of the SIDS energy infrastructure and the lengthy recovery period after suffering (Bundhoo, Shah et al. 2018). To evaluate the resilience and infrastructure dynamics that influence the long-term stability of civil infrastructure systems, Rasoulkhani and Mostafavi (2018) proposed a multi-agent simulation model. The suggested framework takes three factors to frame the dynamics of the interacting human-infrastructure systems, including engineered physical infrastructure, human actors, and chronic and severe stressors. They pointed out that the performance mechanism of infrastructure systems depends on the infrastructure's resilience. These interrelated elements simultaneously affect IR systems by functioning in concert (Rasoulkhani and Mostafavi 2018).

6.7.2 Assessments of resilience in infrastructures

After understanding the contexts and the underlying factors, the next step entails detailed assessments of the impact being exerted on the infrastructure systems. In the assessment process, risks and vulnerability of all the relevant factors including their collective impacts are important to analyse about the infrastructure operations. Dependence and interdependence analysis of the infrastructure systems under various contexts need to be examined by involving relevant actors or stakeholders across the operational processes. The process may result in a comprehensive resilience dependency matrix depicting the functions, roles, and responsibilities of the stakeholders against the infrastructure types across single or multiple operational scenarios.

6.7.3 Qualification and quantification of resilience in infrastructures

Appropriate qualification and quantifications are two key processes in determining preventative steps to lessen the effects of disruptive occurrences and require assessing the resilience of the infrastructure (Singhal, Kwon et al. 2020). Numerous academics have used various analytical techniques to evaluate how IR responds to interruptions. For instance, Levenberg et al. (2017) quantified and evaluated the resilience of networked pavement infrastructure by simulating a variety of potential network performance scenarios in a destructive meteorological scenario with known probabilities of occurrence; each scenario was specified in terms of the degree and type of damage (climate or geology, procedure, natural degradation, and terrorist acts), as well as the current weather, including temperature, precipitation, and visibility. An agent-based modelling approach for the seismic resilience evaluation of integrated civil infrastructure systems under a seismic scenario has been proposed by Sun et al. The robustness of the healthcare infrastructure before and after COVID-19 was evaluated by Barabadi et al. (2020). A multidimensional assessment index method was put up by Xu et al. (2023) to evaluate the infrastructure systems' resilience to the harsh weather in Wuhan.

Due to the virtue of infrastructure operations being highly interdependent, it can be challenging to measure resilience comprising entire infrastructure systems (Carpenter, Walker et al. 2001). A comprehensive assessment framework for resilience was proposed by Bruneau et al. (2003), who noted that 11 aspects, including four dimensions or contexts (technical, organisational, social, and economic), four fundamental properties (robustness, rapidity, redundancy, and resourcefulness), as well as three outcomes, need to be taken into account (i.e. more reliable, faster recovery, lower consequences). To construct catastrophe management, Huck et al. (2020) employed the four R approach, integrating robustness, rapidity, redundancy, and resourcefulness to measure the power system's resilience. To evaluate the resilience of electric infrastructure systems using the four R concept and re-adjustability, Toroghi and Thomas (2020) established a framework with five criteria (robustness, resourcefulness, redundancy, rapidity, and readability).

One of the most crucial characteristics of resilience is robustness, which is the capacity of the system to endure anomalous and dangerous circumstances (Alizadeh and Sharifi 2020). A resilience failure will occur when one or more of the linked system's capabilities are depleted; this event is initially related to the system's robustness (Faber, Miraglia et al. 2018). Numerous pieces of research evaluated the robustness of infrastructure in terms of resilience (Sela, Bhatia et al. 2017, Homayounfar, Muneepeerakul et al. 2018). To evaluate the relationship between resilience and robustness of coupled infrastructure systems, Homayounfar et al. (2018) used a stylised dynamical model, asserting that robustness is an element of resilience. This model simulates how the robustness and resilience of critical infrastructure systems react to shocks in state variables change with parameters.

Redundancy, one of the measures used to gauge a system's resilience, can increase a system's reliability by duplicating its essential parts or operations (Chopra, Dillon et al. 2016). Perz et al. (2012) used a rural sample survey to analyse the correlation between community and town interconnection and socioecological resilience, examining the impact of resilience and interconnection on infrastructure. They used a three-country border in the southwest Amazon connected by a highway as a case study. To examine the strategy for enhancing the redundancy of infrastructure networks,

Capacci and Biondini (2020) established a probabilistic framework for evaluating the life-cycle seismic resilience of infrastructure networks incorporating deteriorating bridges and transportation routes.

6.7.4 *Improvement of the resilience in infrastructures*

A resilient infrastructure system should have four characteristics – robustness, redundancy, speed, and resourcefulness – to reduce the likelihood of failing, have redundant connections, shorten recovery times, and limit impact spread (Sun, Bocchini et al. 2018).

The infrastructure systems should improve their resilience and stability while responding to threats or assaults to reduce the likelihood of failure. The resistance to withstand various dangers may be improved by utilising a variety of approaches, according to several pieces of research (Johansen and Tien 2017, Zhao, Liu et al. 2017). For example, to improve and maximise the resilience of dynamic infrastructure systems, Zhao et al. (2017) created non-homogeneous hidden Markov models for resilience measures in various damage scenarios, including terrorist attacks, human-originated catastrophes, and natural disasters. The Bayesian Network methods have been employed by Johansen and Tien (2017) to find the worst effects and prioritise elements for the repair of reinforcement, assisting and delivering tactics for increasing the resilience of infrastructure systems based on probabilistic modelling and under natural and targeted attacks.

The improvement of urban resiliency through employing green infrastructure systems has been the focus of several researches (Staddon, Ward et al. 2018, Lee, Song et al. 2021). For instance, Staddon et al. (2018) highlighted that green infrastructure systems significantly contribute to increasing urban resilience. They performed a high-level literature review to recognise the benefits of green infrastructure to urban resilience and demonstrated the problems of sustaining this green infrastructure-based approach for identifying methods to improve urban robustness (Staddon, Ward et al. 2018). Based on the flood-adaptive design strategy and simulation results, Lee et al. (2021) proposed a green infrastructure-based strategy for increasing the resilience of urban infrastructure systems to flood. They simulated the effect of reduced stormwater runoff by employing green infrastructure in the Gangnam district of Seoul (Lee, Song et al. 2021).

Under typical operating conditions, the interdependence and interconnectedness of systems benefit each other, as they enhance the capacity of urban infrastructure networks to withstand severe events like human attacks and natural catastrophes (Balakrishnan and Zhang 2020). *Connectivity* is defined in graph theory as the least number of components required to divide the remaining nodes into independent subgraphs (Diestel 2017). In general, a larger number of interconnecting links between two nodes can help to increase the infrastructure network's redundancy and decrease its segregation while also making it more accessible and less isolated (Yang, Ng et al. 2020). As network connectivity increases, the efficiency of the infrastructure network may be enhanced, for example, by boosting redundancy and expanding the capacity of crucial linkages and interconnected nodes (Liu, Xiu et al. 2020).

6.8   Summary

This chapter discusses three interrelated issues, risks, vulnerability, and resilience from the rural community perspective. The chapter started with a generic discussion on types of various risks of rural infrastructure. Considering the community's viewpoints and infrastructure development environment, the risks and underlying management processes were further extended. Diversification of rural business activities of one of the key strategies was discussed. In the vulnerability section, the concept of vulnerability in rural communities is discussed highlighting the difference from the preceding risks-based discussions. Then the sources and extent of vulnerability associated with the rural infrastructure were discussed in detail. Factors exerting vulnerability among rural communities in their dependent relations with other people and infrastructure usage were discussed based not a visual network analysis. The final section about community resiliency is discussed focusing first on the risks and vulnerability events and then on processes of building resilience within the specific community concerned. Various factors affecting community resilience and underlying management strategies are highlighted. A new model for vulnerability assessment and management is proposed as a way forward for building resilience among rural communities holistically. The next chapter will discuss cost planning, budgeting, and funding for infrastructure development for supporting and empowering communities in rural settings.

References

Abrams, A. L., K. Carden, C. Teta and K. Wågsæther (2021). "Water, sanitation, and hygiene vulnerability among rural areas and small towns in South Africa: exploring the role of climate change, marginalization, and inequality." Water **13**(20): 2810. https://doi.org/10.3390/w13202810

Acemoglu, D. and F. Zilibotti (1997). "Was Prometheus unbound by chance? Risk, diversification, and growth." Journal of Political Economy **105**(4): 709–751.

Alizadeh, H. and A. Sharifi (2020). "Assessing resilience of urban critical infrastructure networks: a case study of Ahvaz, Iran." Sustainability **12**(9): 3691. https://doi.org/10.3390/su12093691

Anderson, J. R. (2003). "Risk in rural development: challenges for managers and policy makers." Agricultural Systems **75**: 161–197.

Balakrishnan, S. and Z. Zhang (2020). "Criticality and susceptibility indexes for resilience-based ranking and prioritization of components in interdependent infrastructure networks." Journal of Management in Engineering **36**(4). https://doi.org/10.1061/(ASCE)ME.1943-5479.0000769

Barabadi, A., M. H. Ghiasi, A. N. Qarahasanlou and A. Mottahedi (2020). "A holistic view of health infrastructure resilience before and after COVID-19." Archives of Bone and Joint Surgery **8**(Suppl 1): 262.

Barrios, E. B. (2008). "Infrastructure and rural development: Household perceptions on rural development." Progress in Planning **70**(1): 1–44.

Blaikie, P., T. Cannon, I. Davis and B. Wisner (2003). At risk: Natural hazards, people's vulnerability and disasters, Florence, UNITED STATES, Taylor & Francis Group.

Brown, K. (2014). "Global environmental change I: A social turn for resilience?" Progress in Human Geography **38**(1): 107–117.

Bruneau, M., S. E. Chang, R. T. Eguchi, G. C. Lee, T. D. O'Rourke, A. M. Reinhorn, M. Shinozuka, K. Tierney, W. A. Wallace and D. Von Winterfeldt (2003). "A framework to quantitatively assess and enhance the seismic resilience of communities." Earthquake Spectra **19**(4): 733–752.

Buikstra, E., H. Ross, C. A. King, P. G. Baker, D. Hegney, K. McLachlan and C. Rogers-Clark (2010). "The components of resilience-Perceptions of an Australian rural community." Journal of Community Psychology **38**(8): 975–991.

Bundhoo, Z. M. A., K. U. Shah and D. Surroop (2018). "Climate proofing island energy infrastructure systems: Framing resilience based policy interventions." Utilities Policy **55**: 41–51.

Camacho, C., P. Bower, R. T. Webb and L. Munford (September 2023). "Measurement of community resilience using the Baseline Resilience Indicator for Communities (BRIC) framework: A systematic review." International Journal of Disaster Risk Reduction **95**. doi: 10.1016/j.ijdrr.2023.103870

Capacci, L. and F. Biondini (2020). "Probabilistic life-cycle seismic resilience assessment of aging bridge networks considering infrastructure upgrading." Structure and Infrastructure Engineering **16**(4): 659–675.

Carpenter, S., B. Walker, J. M. Anderies and N. Abel (2001). "From metaphor to measurement: resilience of what to what?" Ecosystems **4**(8): 765–781.

Chapman, C. B. (1997). Project risk management: processes, techniques, and insights / Chris Chapman and Stephen Ward, Wiley.

Chopra, S. S., T. Dillon, M. M. Bilec and V. Khanna (2016). "A network-based framework for assessing infrastructure resilience: a case study of the London metro system." Journal of The Royal Society Interface **13**(118): 20160113.

Das Gupta, M., H. Grandvoinnet and M. Romani (2003). "Fostering community-driven development: what role for the state?" Available at SSRN 636331.

Davies, A. L., R. Streeter, I. T. Lawson, K. H. Roucoux and W. Hiles (2018). "The application of resilience concepts in palaeoecology." The Holocene **28**(9): 1523–1534.

Diestel, R. (2017). Graph theory, Heidelberg, Springer Berlin.

Doloi, H. (2018). "Community-centric model for evaluating social value in projects." Journal of Construction Engineering and Management **144**(5). https://doi.org/10.1061/(ASCE)CO.1943-7862.0001473

Duflo, E. and R. Pande (2007). "Dams*." The Quarterly Journal of Economics **122**(2): 601–646.

Economy, G. C. o. t. and Climate (2016). The sustainable infrastructure imperative: Financing for better growth and development. The 2016 New Climate Economy Report, Washington, DC, The New Climate Economy.

Engler, E., D. Göge and S. Brusch (2018). "ResilienceN–a multi-dimensional challenge for maritime infrastructures." NAŠE MORE: znanstveni časopis za more i pomorstvo **65**(2): 123–129.

Faber, M. H., S. Miraglia, J. Qin and M. G. Stewart (2018). "Bridging resilience and sustainability – decision analysis for design and management of infrastructure systems." Sustainable and Resilient Infrastructure **5**(1–2): 102–124.

Fahad, S. and J. Wang (2020). "Climate change, vulnerability, and its impacts in rural Pakistan: a review." Environ Sci Pollut Res Int **27**(2): 1334–1338.

Farahani, H. and M. Jahansoozi (2022). "Analysis of rural households' resilience to drought in Iran, case study: Bajestan County." International Journal of Disaster Risk Reduction **82**. https://doi.org/10.1016/j.ijdrr.2022.103331

Faturechi, R. and E. Miller-Hooks (2015). "Measuring the performance of transportation infrastructure systems in disasters: a comprehensive review." Journal of Infrastructure Systems **21**(1). https://doi.org/10.1061/(ASCE)IS.1943-555X.0000212

Freshwater, D. (2015). "Vulnerability and resilience: Two dimensions of rurality." Sociologia Ruralis **55**(4): 497–515.

Glaeser, E. L. and J. E. Kohlhase (2004). Cities, regions and the decline of transport costs. Fifty years of regional science. R. J. G. M. Florax and D. A. Plane. Berlin, Heidelberg, Springer Berlin Heidelberg: 197–228.

Grimsey, D. and M. K. Lewis (2002). "Evaluating the risks of public private partnerships for infrastructure projects." International Journal of Project Management **20**(2): 107–118.

Headd, B. (2003). "Redefining business success: Distinguishing between closure and failure." Small Business Economics **21**(1): 51–61.

Hemson, D., M. Meyer and K. Maphunye (2004). "Rural development: The provision of basic infrastructure services." (Position paper, support from National Treasury, January). Human Science Research Council (HSRC) Research Outputs, South Africa. http://hdl.handle.net/20.500.11910/8215

Homayounfar, M., R. Muneepeerakul, J. M. Anderies and C. P. Muneepeerakul (2018). "Linking resilience and robustness and uncovering their trade-offs in coupled infrastructure systems." Earth System Dynamics **9**(4): 1159–1168.

Huck, A., J. Monstadt and P. Driessen (2020). "Building urban and infrastructure resilience through connectivity: An institutional perspective on disaster risk management in Christchurch, New Zealand." Cities **98**. https://doi.org/10.1016/j.cities.2019.102573

IFRC (2012). Disaster and crisis management, International Federation of Red Cross Crescent Societies.

ILO (2005). Mainstreaming poverty alleviation strategies through sustainable rural infrastructure development, Bangkok, International Labour Organisation, Employment Intensive Investment Programme.

Imai, K. S., R. Gaiha and G. Thapa (2015). "Does non-farm sector employment reduce rural poverty and vulnerability? Evidence from Vietnam and India." Journal of Asian Economics **36**: 47–61.

Jamshed, A., J. Birkmann, I. A. Rana and J. M. McMillan (2020). "The relevance of city size to the vulnerability of surrounding rural areas: An empirical study of flooding in Pakistan." International Journal of Disaster Risk Reduction **48**. doi: 10.1016/j.ijdrr.2020.101601

Johansen, C. and I. Tien (2017). "Probabilistic multi-scale modeling of interdependencies between critical infrastructure systems for resilience." Sustainable and Resilient Infrastructure **3**(1): 1–15.

Kelly, C., A. Ferrara, G. A. Wilson, F. Ripullone, A. Nolè, N. Harmer and L. Salvati (2015). "Community resilience and land degradation in forest and shrubland socio-ecological systems: Evidence from Gorgoglione, Basilicata, Italy." Land Use Policy **46**: 11–20.

Keshavarz, M., H. Maleksaeidi and E. Karami (2017). "Livelihood vulnerability to drought: A case of rural Iran." International Journal of Disaster Risk Reduction **21**: 223–230.

Khalid, Z., X. Meng, I. A. Rana, M. u. Rehman and X. Su (2021). "Holistic Multidimensional Vulnerability Assessment: An empirical investigation on rural communities of the Hindu Kush Himalayan region, Northern Pakistan." International Journal of Disaster Risk Reduction **62**. doi: 10.1016/j.ijdrr.2021.102413

Khan, A. M. (2013). "Risk factors in toll road life cycle analysis." Transportmetrica A: Transport Science **9**(5): 408–428.

Kostov, P. and J. Lingard (2001). Rural development as risk management, University of Newcastle upon Tyne, Centre for Rural Economy.

Kostov, P. and J. Lingard (2003). "Risk management: a general framework for rural development." Journal of Rural Studies **19**(4): 463–476.

Kumari, A. and A. Kumar Sharma (2017). "Infrastructure financing and development: A bibliometric review." International Journal of Critical Infrastructure Protection **16**: 49–65.

Lazarte, A. (2017). Understanding the drivers of rural vulnerability: Towards building resilience, promoting socio-economic empowerment and enhancing the socio-economic inclusion of vulnerable, disadvantaged and marginalized populations for an effective promotion of Decent Work in rural economies, Geneva, Switzerland, Employment Policy Department, International Labour Organization.

Lee, H., K. Song, G. Kim and J. Chon (2021). "Flood-adaptive green infrastructure planning for urban resilience." Landscape and Ecological Engineering **17**(4): 427–437.

Levenberg, E., E. Miller-Hooks, A. Asadabadi and R. Faturechi (2017). "Resilience of networked infrastructure with evolving component conditions: Pavement network application." Journal of Computing in Civil Engineering **31**(3): 04016060.

Lewis, T. G. (2019). Critical infrastructure protection in homeland security: defending a networked nation, John Wiley & Sons.

Liu, H. J., P. E. D. Love, J. Zhao, C. Lemckert and K. Muldoon-Smith (2021). "Transport infrastructure asset resilience: Managing government capabilities." Transportation Research Part D: Transport and Environment **100**. https://doi.org/10.1016/j.trd.2021.103072

Liu, Z., C. Xiu and C. Ye (2020). "Improving urban resilience through green infrastructure: an integrated approach for connectivity conservation in the central city of Shenyang, China." Complexity **2020**: 1–15.

Mao, Q. and N. Li (2018). "Assessment of the impact of interdependencies on the resilience of networked critical infrastructure systems." Natural Hazards **93**(1): 315–337.

Matsukawa, T. and O. Habeck (2007). Review of risk mitigation instruments for infrastructure financing and recent trends and developments, World Bank Publications.

McGee, T. K. and E. C. Penning-Rowsell (2022). Routledge handbook of environmental hazards and society, London, Routledge.

Merna, T. and C. Njiru (2002). Financing infrastructure projects, Thomas Telford.

Nijkamp, P. and S. A. Rienstra (1995). "Private sector involvement in financing and operating transport infrastructure." The Annals of Regional Science **29**(2): 221–235.

Ouyang, M., C. Liu and M. Xu (2023). "Value of resilience-based solutions on critical infrastructure protection: Comparing with robustness-based solutions." Reliability Engineering & System Safety **190**. https://doi.org/10.1016/j.ress.2019.106506

Parrott, M. (2020). "10 million animals are hit on our roads each year. Here's how you can help them (and steer clear of them) these holidays." from https://findanexpert. unimelb.edu.au/news/13968-10-million-animals-are-hit-on-our-roads-each-year.- here%E2%80%99s-how-you-can-help-them-(and-steer-clear-of-them)-these-holidays.

Perz, S. G., A. Shenkin, G. Barnes, L. Cabrera, L. A. Carvalho and J. Castillo (2012). "Connectivity and resilience: A multidimensional analysis of infrastructure impacts in the southwestern Amazon." Social Indicators Research **106**(2): 259–285.

Rahimi-Golkhandan, A., B. Aslani and S. Mohebbi (2022). "Predictive resilience of interdependent water and transportation infrastructures: A sociotechnical approach." Socio-Economic Planning Sciences **80**. https://doi.org/10.1016/j.seps.2021.101166

Rahman, H. M. T. and G. M. Hickey (2020). "An analytical framework for assessing context-specific rural livelihood vulnerability." Sustainability 12(14): 5654. https://doi.org/10.3390/su12145654

Rajesh, S., S. Jain, P. Sharma and R. Bhahuguna (2014). "Assessment of inherent vulnerability of rural communities to environmental hazards in Kimsar region of Uttarakhand, India." Environmental Development 12: 16–36.

Rasoulkhani, K. and A. Mostafavi (2018). "Resilience as an emergent property of human-infrastructure dynamics: A multi-agent simulation model for characterizing regime shifts and tipping point behaviors in infrastructure systems." PLoS One 13(11): e0207674.

Rehak, D., P. Senovsky and S. Slivkova (2018). "Resilience of critical infrastructure elements and its main factors." Systems 6(2): 21. https://doi.org/10.3390/systems6020021

Rigg, J. (2006). "Land, farming, livelihoods, and poverty: Rethinking the links in the Rural South." World Development 34(1): 180–202.

Rigg, J., A. Salamanca and M. Parnwell (2012). "Joining the dots of agrarian change in Asia: a 25 year view from Thailand." World Development 40(7): 1469–1481.

Scholz, M. and B. h. Lee (2005). "Constructed wetlands: a review." International Journal of Environmental Studies 62(4): 421–447.

Sela, L., U. Bhatia, J. Zhuang and A. Ganguly (2017). "Resilience strategies for interdependent multiscale lifeline infrastructure networks." Computing in Civil Engineering 2017: 265–272.

Shakou, L. M., J.-L. Wybo, G. Reniers and G. Boustras (2019). "Developing an innovative framework for enhancing the resilience of critical infrastructure to climate change." Safety Science 118: 364–378.

Shi, J., S. Wen, X. Zhao and G. Wu (2019). "Sustainable development of urban rail transit networks: a vulnerability perspective." Sustainability 11(5): 1335. https://doi.org/10.3390/su11051335

Singhal, T. K., O.-S. Kwon, E. C. Bentz and C. Christopoulos (2020). "Development of a civil infrastructure resilience assessment framework and its application to a nuclear power plant." Structure and Infrastructure Engineering 18(1): 1–14.

Smith, N. J., T. Merna and P. Jobling (2014). Managing risk in construction projects, John Wiley & Sons.

Splichalova, A., D. Patrman, N. Kotalova and M. Hromada (2020). "Managerial decision making in indicating a disruption of critical infrastructure element resilience." Administrative Sciences 10(3): 75. https://doi.org/10.3390/admsci10030075

Staddon, C., S. Ward, L. De Vito, A. Zuniga-Teran, A. K. Gerlak, Y. Schoeman, A. Hart and G. Booth (2018). "Contributions of green infrastructure to enhancing urban resilience." Environment Systems and Decisions 38(3): 330–338.

Sun, W., P. Bocchini and B. D. Davison (2018). "Resilience metrics and measurement methods for transportation infrastructure: the state of the art." Sustainable and Resilient Infrastructure 5(3): 168–199.

Thacker, S., D. Adshead, M. Fay, S. Hallegatte, M. Harvey, H. Meller, N. O'Regan, J. Rozenberg, G. Watkins and J. W. Hall (2019). "Infrastructure for sustainable development." Nature Sustainability 2(4): 324–331.

Thobani, M. (1999). "Private infrastructure, public risk." Finance & Development 0036(001): A013.

Tobin, G. A. (1999). "Sustainability and community resilience: the holy grail of hazards planning?" Environmental Hazards 1(1): 13–25.

Toroghi, S. S. H. and V. M. Thomas (2020). "A framework for the resilience analysis of electric infrastructure systems including temporary generation systems." Reliability Engineering & System Safety 202: 107013–107028. https://doi.org/10.1016/j.ress.2020.107013

UNEP (2002). Assessing human vulnerability to environmental change: concepts, issues, methods and case studies, United Nations Environment Programme.

UNISDR (2015). Global Assessment Report on Disaster Risk Reduction 2015. USDOC economic impact of Hurricane Sandy: potential economic activity lost and gained in New Jersey and New York, Economics and Statistics Administration Office of the Chief Economist.

Wilson, G., G. Quaranta, C. Kelly and R. Salvia (2015). "Community resilience, land degradation and endogenous lock-in effects: evidence from the Alento region, Campania, Italy." Journal of Environmental Planning and Management **59**(3): 518–537.

Wilson, G. A. (2012). Community resilience and environmental transitions, Taylor & amp; Francis.

Wilson, G. A. (2012). "Community resilience, globalization, and transitional pathways of decision-making." Geoforum **43**(6): 1218–1231.

Wilson, G. A. (2015). "Community resilience and social memory." Environmental Values **24**(2): 227–257.

Wilson, G. A., Z. Hu and S. Rahman (2018). "Community resilience in rural China: The case of Hu Village, Sichuan Province." Journal of Rural Studies **60**: 130–140.

Wilson, G. A., M. Schermer and R. Stotten (2018). "The resilience and vulnerability of remote mountain communities: The case of Vent, Austrian Alps." Land Use Policy **71**: 372–383.

Xu, W., J. Cong and D. G. Proverbs (2021). "Evaluation of infrastructure resilience." International Journal of Building Pathology and Adaptation **41**(2): 378–400.

Yang, Y., S. T. Ng, S. Zhou, F. J. Xu and H. Li (2020). "Physics-based resilience assessment of interdependent civil infrastructure systems with condition-varying components: A case with stormwater drainage system and road transport system." Sustainable Cities and Society **54**. https://doi.org/10.1016/j.scs.2019.101886

Zhao, S., X. Liu and Y. Zhuo (2017). "Hybrid Hidden Markov Models for resilience metrics in a dynamic infrastructure system." Reliability Engineering & System Safety **164**: 84–97.

7 Cost planning, budgeting, funding, and procurement

7.1 Introduction

While infrastructures provide many essential services to the community and serve as backbone of the society, infrastructures could be too costly and risky to construct especially depending on the size, complexity, and development environment (Thacker, Adshead et al. 2019). As discussed in the earlier chapter, whether a particular infrastructure is categorised as social infrastructure as a public good or economic infrastructure contributing to the national economy, complexity increases by the scale of the economy and socioeconomic status of the community at large. Within the social infrastructure category, there are some essential infrastructure that accessibility for everyone in the community is mandatory. For instance, infrastructures such as flood-protection systems, law-enforcement infrastructures, water purifying infrastructures, national defence infrastructures, etc. are public goods as services of these infrastructures should be available and accessible for all members of the society. Some other social infrastructures such as public leisure facilities, sports, and recreational facilities are not necessarily to be accessed by everyone in the community and thus these could be classified as merit goods rather than public goods. While the services rendered by the merit goods are free and accessible by anyone, usually these facilities are not as ubiquitous as public goods. On the other hand, the services of the economic infrastructures such as roads, bridges, railways, power, water, irrigation, etc. are essential for the society to function and participate in the economic activities leading to general well-being and community empowerment.

Global Infrastructure Outlook reported that in 2015, 2.3 trillion USD was spent on infrastructure development and the average rate of annual growth for the spending was around 2.9% per year it is suggested that the amount of investment in infrastructure could rise by around 94 trillion USD by 2040 (Outlook 2017) (Bhattacharya, Meltzer et al. 2016). Among the Asian countries, the China-Pakistan corridor as a part of the Belt and Road initiative involved 62 billion USD of Chinese investment (Ascensão, Fahrig et al. 2018, Thacker, Adshead et al. 2019). As reported by Saha and Hong (2021), annually 80 billion USD have been invested in the infrastructure

DOI: 10.1201/9781032622323-7

projects of low-income and middle-income countries (LMICs) (Saha, Hong et al. 2021) out of which 24 billion USD per year funded by the World Bank (Thacker, Adshead et al. 2019) or 21.5 billion USD is funded by the European Investment Bank.

Despite the considerable amount of investment in infrastructure, there is still a considerable gap between demand and capacity in terms of infrastructure development especially in the poor parts of the world. Rapid urbanisation, population expansion, and industrialisation of developing economies are the root causes for driving the current wave of huge infrastructure expenditures. The majority of high-income nations have a considerable stock of ageing infrastructure assets that need to be replaced, renovated, or removed, necessitating significant investments even in nations with reasonably advanced and mature infrastructure networks.

7.2   Infrastructure costs for meeting the growing demands

While the infrastructure spending in developed economies is significantly higher compared to the developing countries, appropriate strategies should be devised for guiding the required volume of investments should be determined in context-based scenarios. While in the United States alone, more than 400 billion USD is being spent on infrastructures annually (Brooks and Liscow 2021), there is variation in the country-specific infrastructure investment needs especially in the LMICs may be quite large (Rozenberg and Fay 2019). How much a particular country needs to invest in infrastructure depends on the targets set about the demographic profile of the population, short- and long-term GDP growth trajectory and quality and efficiency of services in the location-specific environment. As asserted by Rozenberg and Fay (2019), investments of about 4.5 of the GDP may be good enough for the LMICs to meet the targets to serve the community with basic needs and fulfil the requirements of the UN's SDGs. Some of the basic needs may include access to potable water, reasonable sanitation, affordable power, food security, reducing vulnerability, managing risks, meeting the targets for decarbonisation, and reducing the carbon footprints eventually for the community to lead a healthy life and flourish.

As the vast majority of the rural communities reside predominately in the LMICs, one of the key issues around the infrastructure investment costs is a precise determination of the volume of infrastructure spending in specific locations or regions. In making infrastructure investment decisions, understanding the significance of an individual infrastructure on the overall infrastructure systems or network is important. For instance, interdependent relations of transport and power infrastructure concerning consumers' demand and affordability about the engagement in the local businesses and enterprises, job prospects, income generation including economic growth trajectories, etc. are important to examine before deciding on the size of investment. The trade-offs in infrastructure spending rely on balancing the requirements for social equity targets, environmental preservations, and political aspirations including market conditions and opportunities.

While numerous approaches exist for making infrastructure spending decisions, "what-if" assessment based on GDP-based variable targets is considered effective for setting up short-, medium-, and long-term targets. Figure 7.1 shows the cost of infrastructure investment ranges from 2% to 8% of GDP per year in LMICs (Rozenberg and Fay 2019). As seen, with a 2% GDP growth target for the combined infrastructure investment in electricity, transport, water supply and sanitation, flood projects, and irrigations, the individual share of investment needs to be assessed on the minimum spending scenario and with less ambitious but the delivery of high-efficiency services. The example of this minimum spending scenario comprises GDP expenditure of 0.9% in electricity, 0.53% in transport, 0.32% in water supply and sanitation, 0.06% in flood protection, and 0.12% in irrigation systems. As a second scenario of slightly ambitious goals and high services efficiency, total GDP growth is set to be 4.5% with underlying individual targets of 2.2%, 1.3%, 0.55%, 0.32%, and 0.13% in the respective infrastructure sectors. For such an ambitious target, every infrastructure sector needs to embrace emerging technologies and perform at its best to meet the collective target. The third scenario which entails maximum spending with high ambition but slightly compromised service efficiency may result in maximum growth at 8% of GDP. The respective performance target across all five sectors

Average annual cost to develop infrastructure for the preferred scenario and full range of results, by sector, 2015–30

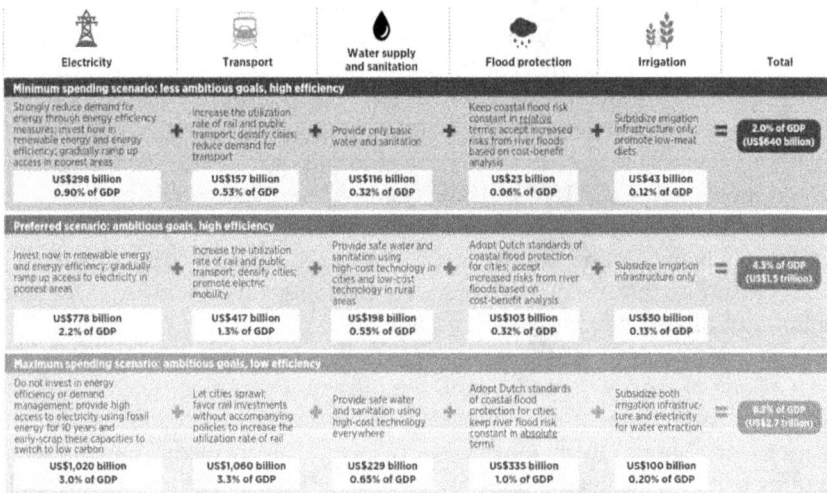

	Electricity	Transport	Water supply and sanitation	Flood protection	Irrigation	Total
Minimum spending scenario: less ambitious goals, high efficiency	Strongly reduce demand for energy through energy efficiency measures; invest now in renewable energy and energy efficiency; gradually ramp up access in poorest areas	Increase the utilization rate of rail and public transport; densify cities; reduce demand for transport	Provide only basic water and sanitation	Keep coastal flood risk constant in relative terms; accept increased risks from river floods based on cost-benefit analysis	Subsidize irrigation infrastructure only; promote low-meat diets	**2.0% of GDP (US$640 billion)**
	US$298 billion 0.90% of GDP	US$157 billion 0.53% of GDP	US$116 billion 0.32% of GDP	US$23 billion 0.06% of GDP	US$43 billion 0.12% of GDP	
Preferred scenario: ambitious goals, high efficiency	Invest now in renewable energy and energy efficiency; gradually ramp up access to electricity in poorest areas	Increase the utilization rate of rail and public transport; densify cities; promote electric mobility	Provide safe water and sanitation using high-cost technology in cities and low-cost technology in rural areas	Adopt Dutch standards of coastal flood protection for cities; accept increased risks from river floods based on cost-benefit analysis	Subsidize irrigation infrastructure only	**4.5% of GDP (US$1.5 trillion)**
	US$778 billion 2.2% of GDP	US$417 billion 1.3% of GDP	US$198 billion 0.55% of GDP	US$103 billion 0.32% of GDP	US$50 billion 0.13% of GDP	
Maximum spending scenario: ambitious goals, low efficiency	Do not invest in energy efficiency or demand management; provide high access to electricity using fossil energy for 10 years and early-scrap these capacities to switch to low carbon	Let cities sprawl; favor rail investments without accompanying policies to increase the utilization rate of rail	Provide safe water and sanitation using high-cost technology everywhere	Adopt Dutch standards of coastal flood protection for cities; keep river flood risk constant in absolute terms	Subsidize both irrigation infrastructure and electricity for water extraction	**8.2% of GDP (US$2.7 trillion)**
	US$1,020 billion 3.0% of GDP	US$1,060 billion 3.3% of GDP	US$229 billion 0.65% of GDP	US$335 billion 1.0% of GDP	US$100 billion 0.20% of GDP	

Figure 7.1 Cost of infrastructure investment ranges from 2% to 8% of GDP per year in low-middle income countries.

Rozenberg, Julie and Fay, Marianne 2019. Beyond the Gap: How Countries Can Afford the Infrastructure They Need while Protecting the Planet (English). Washington, D.C.: World Bank Group. http://documents.worldbank.org/curated/en/189471550755819133/Beyond-the-Gap-How-Countries-Can-Afford-the-Infrastructure-They-Need-while-Protecting-the-Planet will then require to integration high growth strategy including disruptive technologies that result in 3% in electricity, 3.3% in transport, 0.65% in water supply and sanitation, 1.0% in flood protection, and 0.2% in irrigation.

One of the key infrastructure challenges especially in the LMICs is the proportionate distribution of capital and maintenance funds so that infrastructure is utilised to deliver the targeted services and meet the expectations of the community. As much as capital investment is important to increasing the infrastructure stock, upkeeping and maintenance are equally important to render services to the community. Thus, a fine balance needs to be stuck between capital and maintenance costs in all investments to meet the long-term GDP growth as per the target. Figure 7.2 highlights the relative investment costs in Sub-Saharan Africa and South Asia in particular spending scenarios (Rozenberg and Fay 2019). As seen, generally more the investment capital outlay, the more the maintenance costs but the proportion is not always the same across the sectors. For instance, transport is a heavily used physical infrastructure, maintenance costs are higher than in some other sectors such as irrigation or flood protection. Again, the size and proportion of capital versus maintenance costs vary from region to region and country to country. Nonetheless, the most important determinants supporting decisions in each sector of investment are upfront capital outlay and the downstream maintenance regimes so that target performance is achieved.

Average annual cost of investment in the preferred scenario, by sector and region, 2015-30
% of regional GDP

		Africa and Middle East (SSP region)		Asia	Latin America and Caribbean[a] (SSP)	Former Soviet Union[b] (SSP)	
		Middle East and North Africa	Sub-Saharan Africa	South Asia	East Asia and Pacific	Latin America and Caribbean	Eastern Europe and Central Asia[b]
Sector	Type of investment	(World Bank region)					
Electricity	Capital	1.3		2.4		1.2	5.3
	Maintenance	0.3		0.7		0.2	1.1
Transport	Capital	3.2		0.8		1.4	0.0
	Maintenance	1.0		1.6		0.6	1.8
Water supply and sanitation	Capital	0.9	1.6	0.8	0.3	0.5	0.4
	Maintenance	0.3	0.6	0.3	0.1	0.2	0.1
Irrigation	Capital	0.1	0.4	0.3	0.1	0.1	0.0
Flood protection	Capital	0.2	0.8	0.5	0.3	0.2	0.06
	Maintenance	0.04	0.11	0.07	0.08	0.08	0.01
Total[d]	Capital	5.6	7.2	4.8	4.0	3.4	
	Maintenance	1.6	2.0	2.7	2.5	1.1	

Note: Country groups differ between sectors due to the different regional aggregation of models used. SSP = shared socioeconomic pathway, as used by the Intergovernmental Panel on Climate Change.
a. The following countries and territories are included in the SSP country group but not in the World Bank country group: Aruba, The Bahamas, Barbados, Chile, French Guiana, Guadeloupe, Martinique, and Uruguay.
b. The Russian Federation is included in the SSP Former Soviet Union group, but not in the World Bank Eastern Europe and Central Asia group because it is classified as a high-income country.
c. Includes maintenance.
d. Based on countries that are included in all studies.

Figure 7.2 In the preferred scenario, investment costs are the highest for Sub-Saharan Africa and South Asia.

Rozenberg, Julie and Fay, Marianne 2019. Beyond the Gap: How Countries Can Afford the Infrastructure They Need while Protecting the Planet (English). Washington, D.C.: World Bank Group. http://documents.worldbank.org/curated/en/189471550755819133/Beyond-the-Gap-How-Countries-Can-Afford-the-Infrastructure-They-Need-while-Protecting-the-Planet

7.3   Cost of infrastructure and cost overruns

Depending on the growth target, the size of infrastructure spending will vary. In the growth investment scenarios, the complexity of the project increases due to several reasons. For instance, to meet the ambitious targets in the high-growth infrastructure sector, reliance on modern and expensive technology may be necessary. In such a situation, controlling and containing costs from planning to delivery, operations to maintenance is a difficult task and with the increase of project complexity, the difficulty in containing costs increases at an exponential rate.

There are numerous reports of cost overruns in infrastructure projects. In 2016, the Grattan Institute (2016) report highlighted that Australian governments have overspent on transportation infrastructure over the previous 15 years by $28 billion across several inland and interstate projects. That research examined the cost results of all 836 projects of $20 million or more proposed or constructed since 2001. Costs exceeded almost one-fourth of the overall project budgets. It was asserted that about 17% of projects suffer from cost overruns during the construction stage (Terill 2016). Cost overruns being a common phenomenon, especially in large and complex infrastructure projects, the issues seldom garner public notice and often root causes are left unaddressed.

While the scale and extent could be different between developed and developing economies, generally cost overruns are a chronic issue across the globe. As reported in Economic Times (2022), as many as 393 infrastructure projects in India suffered from cost overruns over 78% (Economic Times 25 Sept 2022). As reported by Iyer and Jha (2005), over 40% of infrastructure projects in India are supported by poor performance across the country. Based on the assessments of 951 Indian infrastructure projects, the Ministry of Statistics and Program Implementation (MOSPI) reported that 309 projects suffered from cost overruns while 474 projects suffered from time delays during construction. The report highlighted that over USD 12.4 billion was spent more than the original budget estimate, out of which over USD 8.4 billion overspent amount was due to scheduled delays in projects. Some of the key reasons for cost escalation and schedule delay include poor planning, delay in land acquisition, poor stakeholder coordination, and lack of monitoring and control.

While the above examples highlight the issues mostly in the urban settings, in the case of rural projects, the problems are multifold. Due to the virtue of rural projects being funded by public funds, often projects suffer from a paucity of funding and slow progress. Project delivery responsibility usually resides on the local and district authorities including the involvement of social enterprises, there is not much competition for delivering the project efficiently. There is usually no clear budget and time in sight required for completing the projects with stipulated functionality and services. Due to cost escalation in one project, there are compromises in other projects which eventually result in not being able to deliver the required infrastructure for the community at large. Thus, the root causes of cost escalation in projects are not only important to examine but also crucial to nip in the bud to save scarce resources, especially in rural settings. The following section discusses some of the underlying factors affecting cost performance in projects.

7.4   Factors affecting cost overruns in projects

Factors affecting cost overruns in infrastructure projects is a widely published topic in the mainstream construction and project management literature (Doloi, Sawhney et al. 2012). Thus, a comprehensive account of this issue is considered beyond the scope of this chapter. However, some of the key factors relevant to infrastructure construction from both urban and rural perspectives are discussed briefly.

Premature business cases, lack of planning, and hasty decisions are usually the key reasons for overruns in projects across many countries. Factors such as political pressure, investment lobby, lack of needs, and demand-based options analysis at the concept stage are responsible for downstream cost escalations in projects. In an Australian context, it was observed that during 2001–2016, only 32% of projects were declared ahead of schedule. However, these projects were responsible for 74% of cost overruns (Terill 2016). One of the key reasons behind such poor cost performance was in fact, the premature announcement of projects without working out the extent of lead-up activities including necessary financing approvals, engagement of qualified and relevant construction professionals and planning and controlling of project delivery phases during construction.

Infrastructure projects appear more appealing than they are when promising to build them for less than it ultimately costs. However, when expenses are understated, it is hard for decision-makers to distinguish between excellent and bad ventures. In other words, having a more understandable and reliable figure of reality would make it possible to get the priorities right and spend the money more appropriately (Terill 2016).

Flyvbjerg et al. (2004) reported that the duration of the project in the implementation phase, size of the project, and type of ownership are the most important factors affecting cost overruns. They concluded that lengthy implementation stages and slow planning and execution of significant trans-portation infrastructure projects should worry decision-makers. Simply said, being slow may be very costly. Therefore, every effort should be made to do preparation, planning, authorisation, and ex-ante review before a project owner decides to move forward and create a project to negotiate and resolve issues that would otherwise emerge as implementation delays. Similar to this, the project structure and management must be established and carried out in ways that minimise the potential of interruptions once the choice to create a project has been made. Furthermore, they suggested that the financiers – whether they be taxpayers or private investors – are likely to be severely penalised in terms of cost increases of a level that might endanger project viability if those in charge of the project fail to accomplish this (Flyvbjerg, Skamris Holm et al. 2004).

In terms of project size, Flyvbjerg et al. (2004) reported that larger projects tend to have higher percentage cost escalations than smaller ones for bridges and tunnels. However, this does not seem to be the case for rail and road projects. They added that the risk of cost escalation is considerable for all project sizes and kinds, and their results do not support the claim that larger projects have a greater risk of cost escalation than smaller ones (Flyvbjerg, Skamris Holm et al. 2004). Besides, they mentioned that over time, projects might grow, but this is only noticeable for road construction (Flyvbjerg, Skamris Holm et al. 2004).

The results of the study of Flyvbjerg et al. (2004) do not support the frequently made assumption that public ownership is inherently problematic and that private ownership is a critical factor in efficiency when it comes to managing cost growth. This does not, however, rule out the idea that there may be further factors favouring private ownership over public ownership, such as the notion that private ownership may assist in shielding the average taxpayer against financial burden and limit the number of persons exposed to it. The fundamental issue with cost growth may not be between public and private ownership but rather between a particular form of public ownership, precisely state-owned firms, which lack both the openness and public oversight that being placed in the public sector would imply, and the competitive pressure that is placed in the private sector would result in (Flyvbjerg, Skamris Holm et al. 2004).

Grattan Institute has recommended seven guidelines for the Australian context to eliminate or diminish the cost overrun (Terill 2016). Based on that Governments should only be permitted to invest public funds in transportation infrastructure if a thorough, impartial comparison of like-for-like projects has been made and the supporting business case has been presented to the state or federal legislature. A better public understanding through publishing a summary report of all transport infrastructures funded by the Australian government within the previous quarter as well as a cost-benefit analysis for all proposed infrastructure projects in the previous quarter should be established. Post-completion data is the other important information that the Grattan Institute recommends being published. It is also suggested that when the expected construction cost to a state exceeds $1 billion, Commonwealth, state, or territory administrations be compelled to draught solo legislation for that transportation infrastructure. The Commonwealth should establish model guidelines that states, and territories can adopt or amend, recommending a common strategy for monitoring and managing project risk, including a declaration of seniority when particular criteria might otherwise clash. The Commonwealth should also seek state collaboration to provide new benchmarking data to enhance risk evaluation in new project bids and public accountability. Central agencies should keep project contingency funds separate from project management and formalise the requirements for drawdown.

7.5   Budgeting of infrastructure projects

Budgeting of a project involves cost outlays over the construction lifecycle based on a realistic plan. Good budgeting practice requires careful planning of project scope, accurate understanding of the itemised costs, and project overheads including provisions for contingencies for managing risks and uncertainties covering the project development phases. Some of the key considerations in budgeting infrastructure projects are discussed below:

7.5.1   *Opportunities and challenges*

Infrastructure finance is one of the major strategies for governments to generate employment, stimulate the economy, and contribute to community wellness. However, scarce funds and conflicting priorities often make investment decisions challenging across public jurisdictions. For example, in a survey in the United States conducted by the International City/County Management Association in 2017 with 601 local government respondents, nearly 42% of those surveyed felt that the jurisdiction's infrastructure required additional funding to maintain even baseline maintenance and that the current state of local infrastructure negatively impacted the quality of life in the neighbourhood (Chen and Bartle 2022). However, just 13% of respondents from local governments agree that the jurisdiction's present infrastructure satisfies community requirements and that appropriate financing is accessible to maintain and enhance the assets. Furthermore, 45% of respondents believe that local infrastructure upgrades are possible and that increased infrastructure investment is preferable (Chen and Bartle 2022).

Many factors affect infrastructure financing issues. First, due to increasing population growth and rapid urbanisation, government infrastructure expenditure needs to be readjusted to match supply with demand (Chen and Bartle 2022). For instance, the average number of automobiles per household in the United States has risen from 1.84 in 1984 to 1.96 in 2015 (Sivak 2017). Moreover, the average number of kilometres driven per vehicle has risen from 10641 in 1984 to 13918 in 2015 (Sivak 2017). The same is valid for air transport, freight traffic, and public transportation, all of which are expected to remain (Bartle and Chen 2014). According to the American Society of Civil Engineers (ASCE 2020), there is a $2 trillion, 10-year investment gap, with an $836 billion shortfall for bridge and highway funding.

Services are in high demand for a variety of socioeconomic reasons across different settings. Flood mitigation is becoming more critical as a result of climate change. Shifting supply chains places pressure on transit systems to handle greater freight flow. Technological advances are altering the cars we drive and the infrastructure that supports them (Chen and Bartle 2022). Urban design and transportation systems are changing due to the younger generation's need for more livable metropolitan areas. Infrastructure that uses fewer resources and encourages users to make less wasteful decisions is necessary for environmental sustainability. Additionally, the pressing need for improved social fairness necessitates fair and equitable access to the services offered by public resources (Chen and Bartle 2022).

The future viability of state and municipal infrastructure financing is threatened on the supply side by increasing capital building costs, declining public infrastructure sources of funding, and limited public sector budgets due to growing medical services and pension expenditures (Chen and Bartle 2022). In addition, the National League of Cities' research (NLC 2016) claims that local governments in the United States are under growing pressure to finance infrastructure due to diminishing and insecure federal and state support as well as rising requirements.

Despite these obstacles, there are also potentials in infrastructure investments. "Infrastructure forms a substantial percentage of our shared wealth, and it transcends political boundaries", said Frankel and Wachs (2016). One of the most critical difficulties facing by the modern government is efficiently managing and developing our common assets, which calls for more than just spending more money. The solution to these problems lies in the efficient administration of public resources (Chen and Bartle 2022). Even while the increase in demand is a problem, it shows that more customers are prepared to pay for the services offered. Individuals and enterprises that will profit from the economic growth sparked by infrastructure investment should be ready to provide financial assistance for the numerous projects that pass the trial run of profitability, provided proper financial instruments are used (Chen and Bartle 2022). One crucial objective is a development that increases local prosperity while causing the least amount of traffic and environmental issues. Although requests for more racial and gender equity and environmental sustainability can burden local politicians, these demands are acceptable democratic representations of how voters want their communities to change. Successful leaders will be those elected and appointed authorities who heed these needs and devise plans to satisfy them. The chance to develop income sources for public asset investments that will support long-term, sustainable economic growth is present. These assets may be created using existing financial tools, but they must be appropriately utilised. Thus, project budgeting is clearly a balancing act across numerous conflicting priorities and numerous challenges which need a careful consideration in the policy-making contexts.

7.5.2 *Budgeting and capital planning for infrastructures*

In general, a strategic planning document that outlines a company's actual and planned capital expenditures as well as how the business intends to finance them, is called capital planning. A robust capital budgeting framework ensures that the budget is cost-effectively and coherently aligned with national strategic aims (OECD 2019). According to the OECD Principles of Budgetary Governance (OECD 2019), capital investment plans should be based on an impartial assessment of economic capacity constraints, infrastructure development requirements, and sectoral and social priorities. They also place an emphasis on judicious cost-benefit analysis of such investments, feasibility, proportional project priority, and overall value for money; that investment decisions should be evaluated without regard to the particular funding arrangement; and the creation and execution of a national framework for promoting public investment (OECD 2019).

There are various options for including capital expenditures in the budgeting process, from creating an entirely separate budget to fully integrating them with current expenses (OECD 2019). Each of them has benefits and drawbacks (OECD 2019). While complete integration might boost flexibility, coordination, and planning, separate budgets can prevent mandated spending like entitlements from crowding out discretionary spending like capital expenditure (Posner, Ryu et al. 2009). According to the statistical data represented by OECD (2019), 74% of the contributing countries in the survey reported that they use integrated approach to submit their capital and current spending while the remaining 26% have separated the budgeting and expenses process.

Similarly, there have been advancements in long-term strategic planning, with more than 50% of OECD countries establishing a comprehensive vision for long-term strategic infrastructure that spans all industries (OECD 2019). In some nations, including Norway and Luxembourg, this practice is new (OECD 2019). The reasons for long-term plans vary significantly between nations and are strongly influenced by the economic situation and strategic goals (OECD 2019). In the examined OECD nations, transportation constraints, population changes, and regional growth imbalances are the most typical drivers of strategic infrastructure plans. Creation of shortlists of important projects that may serve as the foundation for "project pipeline planning" and communication is a solid approach that is presently being used by nations like Ireland and Norway (OECD 2019).

Typically, infrastructure projects take a long time to build and operate. Therefore, although the planning and building stages unavoidably use the bulk of resources, it is important to apportion responsibilities for tracking and assessing projects throughout their existence. In order to do this, the majority of nations (69%) have explicit policies in place to ensure that the relevant line ministry or agency evaluates each project's performance (OECD 2019). Out of these, the policy is set and administered by the central government in 31% of the nations studied, and there is a general mandate in 38% of the countries, but it is up to the line department to decide on such policies (OECD 2019).

Capital spending differs from the current expenditure for three reasons (Chen and Bartle 2022): (1) Capital projects are often expensive, and their finance and financing are complicated and frequently sourced from outside sources. (2) Since capital projects sometimes take a long time to construct, the funds must last for more than a single year. Therefore, capital purchases cannot readily be reversed once made. (3) Capital assets are anticipated to last for many years and significantly influence operating budgets in the future. In conclusion, capital expenditures are distinguished from operational expenditures by the amount spent and the durability of capital assets.

Therefore, state and municipal governments across many countries use capital budgets, capital improvement plans, or both to separate capital from current expenditure (Chen and Bartle 2022).

Development of the capital plan and budget can connect the strategic aims of the community and strive for a long-term capital investment with community as a priority (Mikesell 2013). Besides, capital needs and their implications on the operating budget could be coordinated more easily (Mikesell 2013). The creation of the budget and capital plan facilitates cross-sectoral, intra-, and intergovernmental cooperation (Mikesell 2013). It also allows for the spreading out of capital improvement expenses over time, stabilising tax rates and enhancing future financial flexibility (Mikesell 2013). Creating centralised capital plan and budget assists in avoiding costly, unneeded, and redundant capital projects, as well as improve the efficiency and cost-effectiveness of capital expenditure (Mikesell 2013). It also balances conflicting capital requirements against limited fiscal resources, makes cautious capital project finance and financing arrangements, communicates with residents and stakeholders about future investment needs and operational implications, maintains and upgrades public capital assets, and keeps and enhances the community's credit rating (Mikesell 2013). Indian government's central capital fund with interest-free terms is considered to be an effective capital expenditure instrument for the state government to increase capital budgeting in infrastructure projects across the country (India 2023).

7.5.3 *Budgeting method for infrastructure projects*

Budgeting for an infrastructure project, especially during the initial phases of execution, is an integral part of the project development phase. Regardless of the deficiencies of limited data and undiscovered uncertainties and risks, early budgeting is always necessary to give a range of prices for project development in conjunction with the first quality standards specified by the project sponsor (Tas and Yaman 2005). In addition, the analytic and precise cost estimation of project works is critical during the design phase when most activities and considerations having cost consequences must be selected and agreed upon among project stakeholders (Yu 2006). Cost estimating and budgeting needs are iterative processes throughout the project's life cycle, including implementation, supervision and control, and project completion. Cost estimates grow increasingly accurate as the project progresses because more reliable data accompanies them (Xenidis and Stavrakas 2013).

Various budgeting systems are used in construction projects, each with its strategy and approach to predicting the budget for a particular project. Deterministic approaches are the most commonly used, while more complicated techniques such as Case-Based Reasoning (CBR), Regression analysis, and Monte Carlo simulation are frequently used in practice. This section provides a quick review of approaches regarding theoretical methodology, requirements, accuracy, and risk integration.

7.5.3.1   *Deterministic cost estimation*

Many deterministic approaches are being utilised to estimate the cost of a project and budget with sufficient precision, among which the Total Quantity Method (TQM) and Unit Quantity Method (UQM) are two of the most extensively used methods (Xenidis and Stavrakas 2013).

According to UQM, the project is broken down into quantifiable components, often materials, which make up the project cost. First, each item's projected cost per unit quantity is calculated by adding the individual costs of the component parts needed to generate a unit quantity. Next, the anticipated cost per unit quantity is multiplied by the overall required quantities for this item in the project context. Finally, the total cost of all the elements, calculated using the given technique, is added to determine the entire project budget. As an illustration, the price of one cubic metre of reinforced concrete is calculated by adding the prices of the concrete, steel, and labour used to make the particular quantity. The amount of reinforced concrete cubic metres needed for the entire project is multiplied by this cost. Then, the final budget for the entire project is projected by summing the total expenses of all selected elements (such as painting, flooring, etc.), which are evaluated as stated above. For TQM, the application takes a backwards approach. In this example, the total quantity of all goods' separate constituents is approximated and multiplied by their unit cost. Then, the overall budget for the project is calculated by adding the total expenditures for each item.

Aside from their appropriate precision, the ease of the computations makes deterministic approaches popular. Furthermore, regardless of the project size, all computations are constrained to four basic mathematical processes. As a result, simple software or specialised knowledge is required to estimate the budget of any project. Another benefit is that the findings are given straightforwardly, allowing for instant judgement on their reliability and acceptance, as well as updating final estimates in the event of modifications in the values of project items.

The most important drawback distinguishing the deterministic and nondeterministic approaches is that they do not accurately add uncertainty and hazards into calculations. Risk contingencies can be handled as additional expenses to the expected ones, but they can only be changed once a new estimation is undertaken from the beginning. The budget's inaccuracy is exacerbated by the fact that in deterministic methods, risk contingencies are typically introduced as fixed percentages of the overall budget, using previous experience or regulatory requirements, rather than estimating reserves based on the specific project's requirements at hand. Because of this limitation, deterministic approaches are inefficient for proper, precise, and reasonable budget estimation.

7.5.3.2   *Stochastic cost estimation*

The stochastic approaches for budget estimating differ from the deterministic methods as in these methods, uncertainty is added in the computations in the form of probability density functions. Each unclear cost item is assigned a value from a range of values, and the probability of occurrence is derived using these distributions. The risk for a cost item is expressed by the standard deviation of its probability distribution. There are several stochastic approaches for estimating a project's budget (Touran 2003, Öztaş and Ökmen 2004). The most widely used are those based on simulation, specifically Monte Carlo (Xenidis and Stavrakas 2013). This approach generates many alternative results for a particular mathematical model representing a physical situation. The values that can be allocated to the problem's uncertain variables, as well as the model's ultimate output, are provided as probabilistic distributions.

The Monte Carlo simulation technique provides a good evaluation of the systematic uncertainties without affecting the system while it is easy to comprehend and quantifies the risk in the form of probabilistic distributions. However, all simulation techniques including Monte Carlo simulation need large amounts of data to construct a realistic simulation model. This amount of data is most of the time unachievable if exists. Regardless of this, it is likely that the results become affected by the sample errors that existed in the data.

7.5.3.3   *Case-based reasoning (CBR)*

CBR is a prominent project management approach with numerous uses, including budget estimates. It is a way of solving issues based on facts and expertise gained from previous similar circumstances. CBR systems and tools execute the expert judgement process, which consists of observing the essential aspects that characterise the problem, locating these features in prior comparable situations where they have emerged, and anticipating the development of the issue based on similar problem experience (Kim, An et al. 2004). CBR systems use the same procedure in four phases (Kim, An et al. 2004): (a) storing past cases in databases to represent the system's experience; (b) restoring and reusing stored old examples depending on their similarity to the situation at hand; (c) review and adaption of prior instances to the case at hand; and (d) storage of solved cases in databases all contribute to successfully identifying the resemblance between experience and the situation at hand. A study by Huang and Tseng provides a good pros and cons analysis for this approach (Huang and Tseng 2004).

7.5.3.4   *Regression*

Regression analysis models have been employed in a variety of domains, including budget estimates since they are valid statistical techniques for identifying variable dependencies and justifying predictions evolution of a modelled issue (Xenidis and Stavrakas 2013). Although the mathematical model is simple to comprehend and regression analysis may be carried out using readily available and manageable software tools, Kim et al. (2004) have underlined certain drawbacks they found in the literature (Images 7.1 and 7.2).

Image 7.1 A dilapidated public toilet in a primary school in a village.

Image 7.2 A dilapidated ancient worshipping place of historic significance from the "Ahom King" dynasty in Assam.

7.6 Budget management and optimisation

Life cycle costing analysis (LCCA) has attracted much attention as a decision-support tool in various engineering domains, including civil engineering. The efficacy of infrastructure projects must be evaluated while taking into account all benefits and expenses brought about by the projects, using time frames indicative of the projects' actual lifetime. This has been recognised in both study and practice. Figure 7.3 shows lifecycle phases and strategic considerations of costs while planning a typical infrastructure project. Considering these phases and underlying costs and benefits, for instance, an engineering project may only be viable if the life cycle advantages outweigh the associated expenses (Rackwitz 2000). Therefore, it is crucial to evaluate the viability of engineering projects like social infrastructures at every stage of their life cycle, from conception through decommissioning (Nishijima and Faber 2009).

Contrary to most private business enterprises, infrastructures created to promote societal development serve purposes or are in some other manner connected to advantages and/or disadvantages that, over time, outlive the generations who chose to build them (Nishijima and Faber 2009). Therefore, the evaluation of life cycle costs must account for the expenses suggested for future generations to promote sustainable social development, that is, a development that tries to maximise both the goals of our generation and those of future generations (Nishijima and Faber 2009).

Sometimes maintenance budget of a piece of infrastructure is much higher than the original budget estimates for covering overall expenditures incurred during the stipulated delivery of services. This is unquestionably attributable partly to the performance as decision-makers, gauged by their ability to achieve desired budgets while maintaining the functioning of their portfolio of structures (Nishijima and Faber 2009). They could ask for more if the absence of funding has significant repercussions, including user prices from the decreased functionality of transportation systems. This is usually the case in rural infrastructure operations where services rendered to the community are free of cost. Due to this practical restriction, making the best choice that minimises the predicted total cost does not always result in the best budgeting for the project from a societal perspective. This also relates to a social resource allocation strategy that maximises net social benefit. Therefore, optimising the choice and the overall budget by maximising the societal benefit becomes a challenge in the context of optimum social resource allocation (Nishijima and Faber 2009). In this regard, Doloi (2018) proposed a model for quantification of social value which can be used in the trade-offs against the lifecycle cost of the project (Doloi 2018). In this research, Doloi (2018) showed how the perceptions of the community based on their interests, needs, impacts, and potential usage can be converged into social value in a network context. Social Network Analysis (SNA) was used to investigate various social relations and communications among the stakeholders about the key project issues and convert the underlying network characteristics for estimating the social value from a societal perspective.

Nishijima and Faber (2009) also proposed a method to recognise the optimum decisions regarding structure maintenance and allocation of budget. To do so, they tried to maximise the expected net benefit in cases in which net benefit is comprised of operation benefits, allocated budget, user costs, and costs of delay in maintenance due to lack of budget, and the financial costs to be paid (Nishijima and Faber 2009). Another example of budget optimisation in the context of road safety infrastructure countermeasures is the study

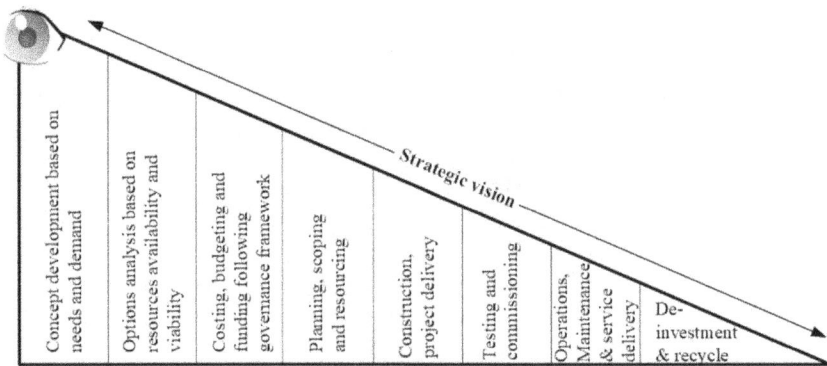

Figure 7.3 Lifecycle phases and strategic cost considerations.

conducted by Byaruhanga and Evdorides (2022) in which the countermeasures are prioritised taking into account their effectiveness throughout their lifecycles together with the constraint of the budget. To do so, they have employed the linear programming technique with an objective function to maximise their safety and economic advantages (Byaruhanga and Evdorides 2022).

Small company structures are crucial to the growth of entrepreneurship and the long-term sustainability of rural communities. In addition, the efficient operation of rural entrepreneurial institutions is necessary for infrastructure reforms and employment. However, small and privately owned farms in rural regions still face the same challenges as most company owners whose operations were impacted by the coronavirus outbreak. Shutting down the borders, disruption of trade and supplies, challenges of emigrating foreign nationals who were engaged in seasonal farm work, and the change in the policy of subsidising and lending to small and medium enterprises are all factors that contributed to the decline of the financial situation and operational conditions (Parushina, Lytneva et al. 2020).

Chronic and unstable financing circumstances especially in the rural infrastructure contexts, rural entrepreneurs operate in an environment of uncertainty and the heightened risks dictate the complexity of the anti-crisis rural business administration policy (Parushina, Lytneva et al. 2020). State and municipal government entities and sectoral ministries and departments are usually responsible for devising context-specific policies, supporting social entrepreneurs to engage over the lifecycle phases of projects and managing the risks and uncertainly appropriately (Parushina, Lytneva et al. 2020).

For social entrepreneurs engaging in rural projects, all financial transfers should be explicitly or implicitly connected to the distribution, allotment, and use of budgetary money to be able to function effectively across the project lifecycle. Thus, financial flows should be clear, transparent, and directed towards attaining the stipulated goals and objectives of the programme's effectiveness indicator (project) (Parushina, Lytneva et al. 2020). The indicator-based performance is attained only when specific budget techniques are executed and exercised with appropriate controlling processes. Both public administrations, private organisations, and social entrepreneurs should be an integral part of the overall management system, which is based on a risk-based approach addressing both external and internal challenges (Parushina, Lytneva et al. 2020).

A budget is frequently cited as the most efficient vehicle for carrying out governmental programmes. The volume of government expenditures and diversity of funding are good indicators of how public projects are prioritised. Government expenditure on rural development should be prioritised across several development areas without undervaluing the contributions of other sectors. For instance, as agriculture continues to be the primary source of income for those living in rural areas, the development of the agricultural sector will directly support the expansion of the rural economy. As frequently stated, improvement in the healthcare and education sectors is crucial for raising Human Development Indexes (HDI). Thus, the prioritisation of the associated infrastructure such as roads, restrooms, and clean water is important for raising the HDIs in rural communities. Funding of this infrastructure and appropriate integration with sectors such as agriculture, education, and health is not only justified but also crucial for rural development (Sutiyo and Maharjan 2017).

7.7 Rural budgeting and associated risks and opportunities

Kiryluk-Dryjska and Beba (2018) provided a strategy for allocating monies for rural development in the European Union based on the categorial measurement of rural development. Scientific rationales such as statistical data or statistical factor analysis were used as underpinning strategies in their funding models (Kiryluk-Dryjska and Beba 2018). Besides, the allocated rural development money is then distributed using a linear programming model. The suggested strategies allocate the funding in a way that favours sectors such as agriculture which were relatively weaker and undeveloped in the specific regions (Kiryluk-Dryjska and Beba 2018). Due to the virtue of decisions being justified on scientific grounds, the approach may also be utilised as a debating tool while negotiating functional allocations in conflicting environments (Kiryluk-Dryjska and Beba 2018).

There are also examples in recent research of how design budget allocation strategies are supported based on several public choice criteria. For instance, emphasising the fairness criteria, Kiryluk-Dryjska (2013) demonstrates how fair division processes can be used for budget distribution. She establishes that these methods increase recipients' acceptance of the initiatives and lessen stakeholder disagreement. Utilising Multiple Criteria Decision Analysis (MCDA) is another approach to this issue. MCDA investigates trade-offs, offers a set of Pareto-efficient solutions, and offers insights into the issue structure. MCDA may be used in politics to rationalise decision-making and lessen stakeholder disputes (Matsatsinis and Samaras 2001). Tinbergen (1952) suggested the idea in 1952, and later same concept was used by many researchers including Chiang (1984). The evolution of MCDA later resulted in computer-based sophisticated modelling and analysis where both quantitative and qualitative criteria are combined in the trade-offs (Estrada and Yap 2013). However such a method is predominately confined to the private sector projects only but with a few exceptions in the public sector projects (Dyer, Fishburn et al. 1992). As a result, case studies and potential applications are always needed to prove the efficacies of these approaches in practice.

According to Dyer et al. (1992), different objective models can help Finland's public sector operate more effectively. Furthermore, a number of researchers such as De Agostini (2006), and Stewart et al. (2009) discuss the use of multi-criteria optimisation to tackle public sector challenges involving access to natural resources. Kim (2007) evaluated the effectiveness of the technique used to examine the Brazilian city of Porto Alegre's financial procedure. Furthermore, Kirschke and Jechlitschka (2002, 2003) suggested linear interactive parametric programming to aid in structural budgeting reforms. Finally, to model the distribution of funding from the European Agricultural Fund for Rural Development in Saxony-Anhalt, Schmid et al. (2010) utilised a real-world example.

Kiryluk-Dryjska (2014), who proposed country-level budgeting, tried to implement linear programming techniques for structural policy (Kiryluk-Dryjska and Beba 2018). However, each region in Poland has been described as having diverse natural, socioeconomic, cultural, and technical characteristics (Kiryluk-Dryjska and Beba 2018). Therefore, allocative decision-making should have been taken into account. In this regard, Kiryluk-Dryjska and Beba (2018) have proposed a region-specific budgeting strategy for rural development. The findings asserted that due to climate change and increasing awareness of sustainable development practices, agri-environment and climate-based policies prevail in the funding allocations in rural regions. The diversity of funding models for attaining the sustainable development goals (SDGs) in localised settings has been underlined. However, from the perspective of using the participatory budget as a tool to achieve SDGs, sustainable rural development is still a mostly unexplored subject. Few nations have implemented laws allowing for the formation of participatory budgets (mainly in rural regions). Therefore, a thorough examination of Poland's experience over ten years may offer crucial recommendations for nations considering doing the same. In this regard, Bednarska-Olejniczak et al. (2020) provide a good explanation of how this concept could support achieving SDGs in rural regions with an emphasis on what can be learnt from the context of Poland.

7.8   Methods for financing infrastructure projects

Every infrastructure project requires investment from a credible funding source. Substantial funding commitment is required for infrastructure projects' lifecycle from the concept design to demolition. However, lack of funding for infrastructure is a widespread issue which is one of the key discussion topics across many countries (Erol and Ozuturk 2011). Traditionally, governments were the primary investors and the sole organisations in charge of building all the infrastructure sectors until the late 1990s (Kumari and Kumar Sharma 2017). However, the sole government's financing from public funds was considered insufficient for addressing the exponential need of infrastructure projects especially in the emerging nations (Kumari and Kumar Sharma 2017). While the decentralised funding and diversified flow of money about the public schemes are considered effective especially for rural infrastructure projects in developing countries, due to inefficacies in the governance structure, a substantial portion of funds gets absolved in operational costs, overheads including salaries of the public officers (Sutiyo and Maharjan 2017). Reforming the governance structure to suit the decentralised funding models is necessary which will be discussed with examples in the following section.

Beginning in the late 1990s, when the private sector started to join the field with substantial finance and advanced operational capabilities, the uptake of projects by private parties under sole financing responsibility was not widespread (Grimsey and Lewis 2002, Agrawal, Gupta et al. 2011). While various models exist about private sector involvement in urban projects, there is not much information on how such models could be extended to rural settings. Some of the commonly used methods for engaging private sectors in urban projects are briefly discussed below (Leruth 2009, Olusola Babatunde, Opawole et al. 2012). However, the key argument of this section intends to be about the ways of engaging private investors and joining hands with public agencies to finance rural infrastructures and meet the needs of rural communities.

7.8.1   *Public investment*

The sole contributor to public infrastructure especially in rural and regional areas is the government. The government had full responsibility for infrastructure development, implementation, operation, maintenance, and repair (Finkenzeller, Dechant et al. 2010). The main objectives of public investments are infrastructure development, environmental protection, employment creation, provision of more excellent living conditions, higher income, and economic development (Kumari and Kumar Sharma 2017). These investments advance society as a whole, not just one single sector but multiple interdependent sectors (Wojewnik-Filipkowska and Trojanowski 2013). Sources of funding for government are usually funds raised from taxes or government bonds and thus rationalisation of public investment is not necessarily to be similar or the same as a private investment. The rationalisation of public investment is usually based on tangible or intangible benefits from the community perspective rather than financial criteria such as return on investment (ROI) (Wojewnik-Filipkowska and Trojanowski 2013).

7.8.2 Private investment

The shift of infrastructure development obligations from the public to the private sector is referred to as private investment (Grimsey and Lewis 2004). The role of private funding in the infrastructure sector is quite substantial across many countries (Kumari and Kumar Sharma 2017). Due to the volume of private funding being available, private investments in meeting the rising infrastructure demands and rendering the necessary services are readily accessible across all sectors in the economies.

Traditionally, the driving force of private investments is growth and financial returns in the infrastructure industry. Thus the investment undergoes stringent assessments of viability, risks and opportunity, market conditions, operating environment, etc. (Sharma and Vohra 2008, Sader 2000). Private investments also bring additional capacity and efficiencies in terms of the use of contemporary technology, improved operational techniques, efficient risk reduction measures, and better planning and timely delivery of the project facilities (Grimsey and Lewis 2002, Gemson, Gautami et al. 2012). While the urban environment with higher user demand is highly conducive to private investments, similar dynamics usually do not exist in rural or regional areas. Due to smaller economies of scale, private investments in rural regions are considered less viable which results in a paucity of funds for rendering the basic services in the rural community. While alternative service delivery, private finance initiatives, public-private partnerships, and foreign direct investment (FDI) are some of the widely used instruments in private sector funding, how these instruments can be used in rural projects is still a question in my economy.

7.8.3 Foreign direct investment (FDI)

When foreign entities make money accessible, but only in situations where the investors actively manage their assets, this is referred to as FDI (Rath and Samal 2015). Many types of FDI are highlighted by Sharma and Vohra (2008) and include divestitures, concessionaire agreements, joint ventures, and greenfield initiatives. Most emerging countries lacked the necessary infrastructure because of a lack of domestic investment. Therefore, these countries need financial aid from other nations, much of which comes from FDIs (Kirkpatrick, Parker et al. 2004, Rath and Samal 2015).

In industries requiring infrastructure, such as power, transportation, telecommunications, health, and education, FDI is crucial (Alsan, Bloom et al. 2006, Sharma and Vohra 2008). Infrastructure and FDI are two sides of the same coin. However, FDI aids in improving infrastructure, and robust infrastructure is essential to luring in additional FDI (Kirkpatrick, Parker et al. 2004).

7.8.4 Public-private partnerships (PPP)

The topic of PPP is widely published in the literature (Grimsey and Lewis 2002, Akintoye, Beck et al. 2008, Agrawal, Gupta et al. 2011). According to

Kumari and Sharma (2017), the essential features of a PPP are that private money helps the creation, management, upkeep, and general growth of infrastructure sectors. In a PPP arrangement, public and private sectors enter into a contract to develop the infrastructure projects and share collective responsibilities for offering necessary services in society.

PPP can take many different shapes (Kumari and Kumar Sharma 2017), including leases, joint ventures, design-bid-build (DBB), design-build (DB), design-build-operate (DBO), design-build-finance-operate (DBFO), and simple operation and maintenance agreements (Sader 2000, Grimsey and Lewis 2002, Merna and Njiru 2002, Grimsey and Lewis 2007).

PPP are essential for several infrastructure sectors, including transportation (Cingolani 2010, Galilea and Medda 2010), energy (Pongsiri 2003), healthcare (Widdus 2017), social housing, waste management (Sindane 2000), education (Utt 1999), water and sanitation, and waste-water treatment (Chiu and Bosher 2005). In PPP arrangements, numerous parties are involved in the collaborations, including credit intermediaries, contractors, special purpose vehicles (SPVs), and organisations from the public and private sectors. As mentioned above, public sector organisations aim to offer services to the public at a reasonable price if not no cost at all, whereas private sector organisations with a profit motivation receive a fee by charging for the services to the public. In PPP arrangements, however, the fee allowed to be charged to the public is controlled by the government with an agreement for a minimum guaranteed ROIs for the investors. To maintain the project-specific partnering arrangements including the design, build, and operation of infrastructure facilities, an SPV is created in a project-specific context (Kumari and Kumar Sharma 2017, Devan 2012).

PPP are a widely accepted option for making infrastructure projects to be efficiently designed, planned, built, and maintained to offer timely and stipulated services to the public without any funding issues (Li, Akintoye et al. 2005, Devan 2012, Hwang, Zhao et al. 2013). However, developing rural projects through PPP presents a significant challenge due to the inability to impose necessary usage charges on the public and meet the expected ROI for the private investors similar to the urban projects.

Over the past decades, there has been an increasing trend of engaging private investments in rural projects across developing countries (Dethier and Moore 2012). With over 40% rural population still residing in rural conditions, more than a quarter live without electricity, telephone, clean drinking water, sanitation and a functional road network. The data from the World Bank suggests that the average private participation in rural projects is just about 13% a year and it's confined to just the top six countries including the Middle East and East Asia (Estache 2010). Public funding share in African projects accounts for about 10–15% but is mainly confined to South Africa and Kenya. Private participation in East Asia, Latin America, and Sub-Saharan Africa is relatively low but increasing steadily (Dethier and Moore 2012).

As stated earlier, the key test of private investment is usually ROI and capital efficiencies, yet for the projects in rural settings, this test does not apply due smaller scale of the economy, lack of productivity, and invariably large capital outlays. Many public projects in rural settings are politically motivated and investment decisions are often not market-tested. Yet, private investments are necessary in projects to create the built-up facilities as public goods and the ROI is usually measured in societal benefits in both tangible and intangible outcomes. One of the prevailing practices of private organisations involved in rural projects is through management contracting where the entire project is funded by the public fund but built and managed by private enterprises. The following sections will discuss some of the common financing practices in developing projects using public and private partnerships.

7.9   Innovation in financing infrastructure projects

Governments in the United States at all three levels have used a range of creative financing techniques to supplement the traditional infrastructure financing strategies in response to the dwindling financial resources available for infrastructure development. However, creative financing has no established definition (Chen and Bartle 2022). Chen (2016) defined this as a broad concept that encompasses all tactics including new resources of funding, new mechanisms of funding, as well as new financial arrangements all of which are a supplement to infrastructure funding sources. Based on the definition of Chen (2016), Chen and Bartle (2022) categorised the innovative infrastructure concept into three types, i.e. new funding sources, new financing mechanisms, and new financial arrangements.

Any new policy that increases money to support infrastructure projects is referred to as a new funding source (Chen and Bartle 2022). This may consist of new taxes, such as gross receipts taxes imposed on villages, municipalities, or companies that directly or indirectly profit from public infrastructure investments or planning choices on a private level. Money raised from such mechanisms may be set aside for infrastructure projects to be spent as both capital outlay or maintenance funds as per the requirements. The new financing mechanisms involve fresh alternatives to borrow money that are flexible and/or could be more affordable (Chen and Bartle 2022). New financial support options such as debt financing, loan guarantees, credit lines, and public and environmental impact bonds form the key sources of funding initiated by the government (Chen and Bartle 2022). The third category, new financial arrangements, involves private, non-profit, or public sectors as new partners to contribute to the financing of the infrastructure and delivery of the project.

Engaging private sector funding into public sector projects requires value judgement to ascertain whether the "value for money" can the ensured from the community's viewpoints. Private sector funds are usually more expensive the public sector funds. The overall cost involved in projects in the PPP procurement arrangements is necessary to assess using an established instrument known as a public-sector comparator (PSC). PSC entails risk-adjustment assessments of the cost of projects in terms of "Net Present Value" by taking into consideration the viewpoints from both public sector vs private sector involvements. In other words, the Net Present Value is calculated in both scenarios to compare and decide whether or not private involvement in a particular project with stipulated terms and conditions is any better than the option of the same project being delivered by the public sector under a similar operating environment. One of the critical factors in such a comparative analysis is the "risks transfer" vs "risks retention" and the costs associated with the uncertain events over the project lifecycle (Estrada and Yap 2013, Podger, Su et al. 2018).

Another widely used measure for assessing the viability of private funds in public projects is the weighted average cost of capital (WACC) which entails the comparison of costs from both public and private sources considering the project phases from planning, construction and operation phases. While WACC is somewhat a measure similar to the ROI measure in a private investment context, WACC usually considers risk-based costing scenarios by incorporating the country or location-specific variabilities. Meaning, that a poorer country with higher inflation rates could result in a much higher WACC compared to a richer country with a lower rate of inflation. Both IRR and WACC indicators eventually represent the cost of private funds in public projects. If these indicators suggest higher costs compared to the funds being used from a public pool, then the use of private funds needs an additional justification such as risks and uncertainties including controlling and mitigation strategies, etc.

7.10 Sources of project financing for rural infrastructures

There are numerous funding sources for rural infrastructure projects. Some of the commonly used funding sources include direct government investment, social enterprise-based funding, or PPP funding (Osei-Kyei and Chan 2015). The sources of funding for government investments are usually concessional loans from development banks and bilateral donors. World Bank is one of the major sources of funding for rural development projects across many developing countries, which affiliates with two major lending agencies, the International Development Agencies (IDA) and the International Bank for Reconstruction and Development (IBRD). Depending on the loan terms of these lending agencies, various countries borrow funds for their development projects at varied concession terms at nominal charges. There are a myriad of other credit agencies for providing country-based and project-specific funding. The key agencies include International Monetary Funds (IMF), African Development Bank, Nordic Development Fund, Asian Development Bank (ADB), Islamic Development Bank, European Union (EU), Organization of the Petroleum Exporting Countries (OPEC), The International Fund for Agricultural

Development (IFAD), Japan Bank for International Cooperation (JBIC), Japan International Cooperation Agency (JICA), The Arab Bank for Economic Development in Africa (BADEA), The European Investment Bank (EIB), etc.

Due to the significant investment need and low profitability of rural infrastructure, its number and quality are inferior to those of urban infrastructure. Therefore, with the proper guiding principles, significant infrastructure development sponsored by the public sector and international organisations will probably result in greater social returns. Among the available options, PPP are efficient investment and management platforms that draw private capital (Wang, Tiong et al. 1999). Cutting-edge financial tools like diversified financing methods also entice social capital to fund infrastructure (Banerjee, Oetzel et al. 2006). PPP models, however, typically are unable to offer financial rewards or risk exposures that satisfy the minimal standards for private sector participation. Therefore, establishing innovative financial and regulatory methods can encourage social capital to increase infrastructure investment, lower the cost of government debt, and provide a broader range of infrastructure and services to society. These methods can also improve the financial environment for social capital to enter the infrastructure field (Jiang, Zhang et al. 2022) (Image 7.3).

Figure 7.4 shows some of the typical sources of funding for developing projects in India. As seen, the two major categories of funding sources are (a) government funds and (b) non-government funds. Government funds are of two types, national and international funds. The national funds are through the national banking institutions and other public lending and credit agencies. International funds are supported by a myriad of international agencies as shown. International agencies usually provide the funds for

Image 7.3 A typical unfit-for-purpose residential home built from a public funding scheme in a village in India.

```
                        ┌──────────────────┐
                        │    Funds for     │
                        │  Development     │
                        └──────────────────┘
              ┌──────────────────┴──────────────────┐
    ┌──────────────────┐                  ┌──────────────────┐
    │ Government Funds  │                  │ Non-government   │
    │                  │                  │     Funds        │
    └──────────────────┘                  └──────────────────┘
```

National funds
- Reserve Bank
- National Banks
- Credit Unions
- Bonds and Securities
- Taxes & Levies

Institutional or non-Institutional Funds
- NGOs
- Private Organisations
- Community organisations
- Farmer's corporation
- Individual sources

International funds
- World Bank
- International Monetary Funds (IMF)
- Organization of the Petroleum Exporting Countries (OPEC)
- Japan International Cooperation Agency (JICA)
- Asian Development Bank (ADB)
- Other development agencies

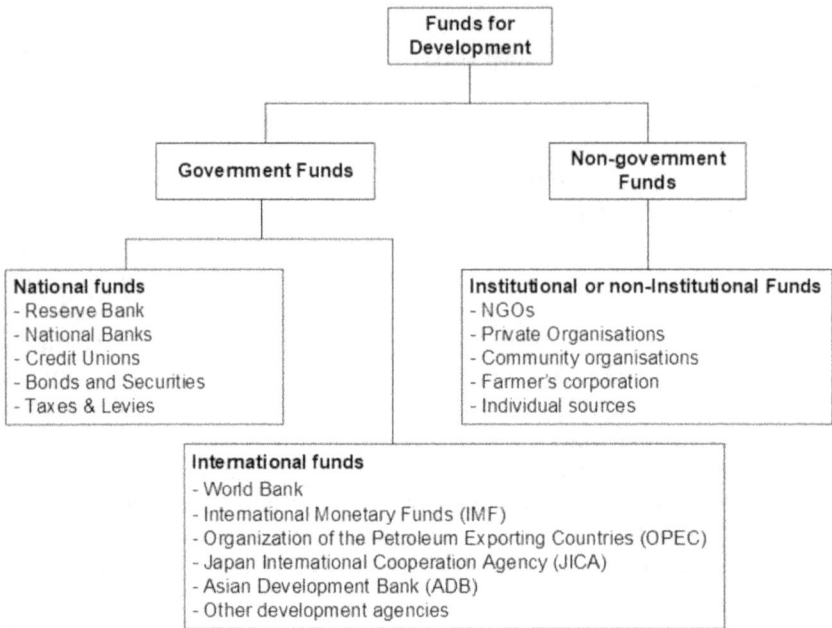

Figure 7.4 Typical sources for funding for developing projects in India.

development projects as low-cost or no-interest loans through the governments. The government is responsible for such funds to administer as well as repay following the stipulated lending terms.

Non-government funds can be either institutional or non-institutional. The sources of such funding include non-government organisations (NGOs), private organisations, community organisations, farmer's corporations individual donors, etc. Non-government funds are usually available for project-based development and are administered privately.

7.11   Procurement strategies for rural infrastructure projects

Public procurement is a key instrument for ensuring the delivery of services by the government to the public and by using scarce resources effectively. Due to the high volume of procurement budgets across numerous interconnected sectors, a well-established governance model is crucial for effective management of project delivery and maintaining accountability and transparency. Procurement strategy plays a significant role in addressing strategic objectives and achieving the end goals in national and international contexts. While procurement strategy is a well-published topic in the mainstream literature including myriad publications by the leading international bodies such as the Organization for Economic Cooperation and Development

(OECD) or Work Bank, there are numerous issues faced by the countries in procurement practices. As project procurement could be a full chapter or even a full book, aligning the scope of the current book, the context-specific discussion of the topic of the delivery of infrastructure projects and underlying governance in rural development is included throughout the book.

In developing a project, procurement processes are involved in two phases, project-level procurement and services procurement within the project delivery context. Procurement at the project level is closely linked through the procurement contracts and underlying contract administrations. Procurement of services entails processes involved in delivering the projects encompassing planning, construction, and operations phases. Research conducted by Benamghar and Iimi (2011) highlights several policy issues in public road procurement. Examining the data from 155 rural road improvement contracts in Nepal, the research posited five key questions in rural road projects (Benamghar and Iimi 2011). First, in competitive bidding for rural road contracts, how do bidders choose their bidding strategy? Second, how do they determine whether or not to enter the procurement market? Third, what are the primary factors of productivity in road construction contracts? Fourth, why do ex-post-contract adjustments occur, such as cost overruns and delays? Finally, which factors are critical or responsible for ensuring the quality of roads produced by the project?

According to the study of Benamghar and Iimi (2011), lack of competition, security issues, and severe weather conditions are the general reasons associated with procurement costs in rural Nepal. Besides, they reported that tender announcements should be extensively publicised and inexpensively distributed to generate more attention and bids, leading to reduced entrance costs for the Firms (2011). They also suggested that national newspapers, rather than local ones, e-bidding or e-procurement can disseminate tender information more broadly and quickly to generate necessary competition for achieving the best outcomes in the project (Benamghar and Iimi 2011).

In a study conducted by Munyede and Mapuva (2020) for the rural areas in Zimbabwe, it is reported that the centralised system of procurement for rural infrastructures is prone to corruption, bureaucratic and is not efficient. According to Mapuva and Miti (2019), countries have succeeded in developing more productive and efficient governance at local levels through decentralisation by growing capacities beyond the sphere of national authority. According to Crawford and Hartmann (2008), decentralisation is the transition of authority, duties, and funds from the central government to sub-national administrations. Another group of scholars contends that decentralisation is a broad notion that may take numerous forms, such as devolution, privatisation, delegation, or deconcentrating (Munyede and Mapuva 2020). As a result, Mapuva and Miti (2019) indicated that each form has distinct traits and qualities. They emphasised that, in practice, devolution has been more successful in federal states such as Germany, Nigeria, and South Africa than in unitary governments.

According to Levine and Bland (2010), devolution is the increased reliance on provincial and district levels of administration with some degree of political autonomy that is significantly beyond direct governmental power but is subject to general policies and regulations affecting the entire country, such as the rule of law. Faguet (2012), on the other hand, believes that devolution is the function that develops or enhances provincial and district divisions of administration, the activities of which are considered outside the jurisdiction of the central government. He claims municipal governments are independent and autonomous, with sole power over clearly reserved responsibilities.

A legislative architecture must be in place to operate as the agora where procurement activities are carried out within the set constraints to provide for regulations regulating how public procurement must be carried out (Munyede and Mapuva 2020). As a result, the OECD (2009, Munyede and Mapuva 2020) suggested that regulatory changes are vital for national economies since they guide market freedom, clarity, cost reductions, and competitiveness. Furthermore, the OECD asserted that policy change is a continual process involving all stakeholders' participation. Natural Resources Governance Institute agreed, stating that the best-designed law is based on extensive engagement and involvement of stakeholders, with a feedback system in place to address any difficulties (Munyede and Mapuva 2020). As seen, project procurement is an integral part of project governance which is further discussed in the next chapter.

7.12   Summary

This chapter discussed the costs and challenges involved in the development of infrastructure projects including meeting the increasing demands. Money being a scarce resource, capital-intensive infrastructure projects have been reviewed from a cost and budget perspective. As cost escalations are a common phenomenon in many infrastructure projects across the world, factors and causality including mitigation strategies have been highlighted briefly. Provisions of budget, cost planning methods, budget management processes, and associated risks and opportunities are discussed from the rural infrastructure perspectives. Then the discussion was further extended on infrastructure finance and various financing methods. Literature-based critical analysis has been included for assessing the pros and cons of some of the widely used financial methods from the decision-making perspective. The significance of innovative financing along with various methods and sources of project funding was further highlighted in developing rural projects, especially in developing countries. Finally, a discussion on project procurement and underlying procurement strategies and the role of project governance for rural development projects are included in the end section of the chapter. The next chapter will expand the discussion on project governance in both theoretical and evidence-based practices and establish a clear framework for achieving target goals for empowering the community through rural infrastructure projects.

References

Agostini, P. d. (2006). "Identifying the best combination of environmental functions using multi-criteria analysis." In J. Cooper, F. Perali and M. Veronesi (eds.), Integrated Assessment and Management of Public Resources: 121–158. Chapter 5, Edward Elgar Publishing.

Agrawal, R., A. Gupta and M. Gupta (2011). "Financing of PPP infrastructure projects in India: constraints and recommendations." IUP Journal of Infrastructure 9(1): 52.

Akintoye, A., M. Beck and C. Hardcastle (2008). Public-private partnerships: managing risks and opportunities, John Wiley & Sons.

Alsan, M., D. E. Bloom and D. Canning (2006). "The effect of population health on foreign direct investment inflows to low- and middle-income countries." World Development 34(4): 613–630.

ASCE (2020). Infrastructure report card, American Society of Civil Engineers.

Ascensão, F., L. Fahrig, A. P. Clevenger, R. T. Corlett, J. A. G. Jaeger, W. F. Laurance and H. M. Pereira (2018). "Environmental challenges for the Belt and Road Initiative." Nature Sustainability 1(5): 206–209.

Banerjee, S. G., J. M. Oetzel and R. Ranganathan (2006). "Private provision of infrastructure in emerging markets: do institutions matter?" Development Policy Review 24(2): 175–202.

Bartle, J. R. and C. Chen (2014). "Future issues in state transportation finance." In M. M. Rubin and K. G. Willoughby (eds.), Sustaining the States: The Fiscal Viability of American State Governments (pp. 211–234). New York, NY: CRC Press.

Bednarska-Olejniczak, D., J. Olejniczak and L. Svobodová (2020). "How a participatory budget can support sustainable rural development—lessons from Poland." Sustainability 12(7): 2620. https://doi.org/10.3390/su12072620

Benamghar, R. and A. Iimi (2011). Efficiency in public procurement in rural road projects of Nepal. World Bank Policy Research Working Paper No. 5736, World Bank.

Bhattacharya, A., J. P. Meltzer, J. Oppenheim, Z. Qureshi and N. Stern (2016). Delivering on sustainable infrastructure for better development and better climate, Gratham Research Institue on Climate Change and the Environment.

Brooks, L. and Z. D. Liscow (2021). "Infrastructure costs." American Economic Journal: Applied (forthcoming).

Byaruhanga, C. B. and H. Evdorides (2022). "A budget optimisation model for road safety infrastructure countermeasures." Cogent Engineering 9(1). https://doi.org/1 0.1080/23311916.2022.2129363

Chen, C. (2016). Innovative infrastructure financing tools. In D. L. Smith and J. B. Justice (eds.), Encyclopedia of public administration and policy. Taylor & Francis Group Press.

Chen, C. and J. R. Bartle (2022). Innovative infrastructure finance, Cham, Palgrave Macmillan.

Chiang, A. (1984). "C., 1984, Fundamental methods of mathematical economics." 3rd ed. Singapore, McGraw-Hill 1984.

Chiu, T. and C. Bosher (2005). Risk sharing in various public private partnership (PPP) arrangements for the provision of water and wastewater services. Conference on Public private Partnerships-Opportunities and Challenges.

Cingolani, M. (2010). "PPP financing in the road sector: a disequilibrium analysis based on the monetary circuit." Transition Studies Review 17(3): 513–550.

Crawford, G. and C. Hartmann (2008). Decentralisation in Africa: a pathway out of poverty and conflict?, Amsterdam University Press.

Dethier, J.-J. and A. Moore (2012). Infrastructure in developing countries: an overview of some economic issues. Discussion Papers on Development Policy No. 165.

Devan, D. (2012). Public private partnerships–Risk management in engineering infrastructure projects, University of Johannesburg (South Africa).

Doloi, H. (2018). "Community-centric model for evaluating social value in projects." Journal of Construction Engineering and Management **5**(144). https://doi.org/10.1 061/(ASCE)CO.1943-7862.0001473

Doloi, H., A. Sawhney, K. C. Iyer and S. Rentala (2012). "Analysing factors affecting delays in Indian construction projects." International Journal of Project Management **30**: 10.

Dyer, J. S., P. C. Fishburn, R. E. Steuer, J. Wallenius and S. Zionts (1992). "Multiple criteria decision making, multiattribute utility theory: the next ten years." Management Science **38**(5): 645–654.

Erol, T. and D. D. Ozuturk (2011). "An alternative model of infrastructure financing based on capital markets: infrastructure REITs (InfraREITs) in Turkey." Journal of Economic Cooperation & Development **32**(3): 65–88.

Estache, A. (2010). "Infrastructure finance in developing countries: An overview." Luxembourg, European Investment Bank (EIB) **15** (2): 60–88.

Estrada, M. A. R. and S. F. Yap (2013). "The origins and evolution of policy modeling." Journal of Policy Modeling **35**(1): 170–182.

Faguet, J.-P. (2012). Decentralization and popular democracy: governance from below in Bolivia, Ann Arbor, USA, University of Michigan Press.

Finkenzeller, K., T. Dechant and W. Schäfers (2010). "Infrastructure: a new dimension of real estate? An asset allocation analysis." Journal of Property Investment & Finance **28**(4): 263–274.

Flyvbjerg, B., M. K. Skamris Holm and S. L. Buhl (2004). "What causes cost overrun in transport infrastructure projects?" Transport Reviews **24**(1): 3–18.

Frankel, E. and M. Wachs (2016). "More than money." Public Works Management & Policy **22**(1): 6–11.

Galilea, P. and F. Medda (2010). "Does the political and economic context influence the success of a transport project? An analysis of transport public-private partnerships." Research in Transportation Economics **30**(1): 102–109.

Gemson, J., K. V. Gautami and A. Thillai Rajan (2012). "Impact of private equity investments in infrastructure projects." Utilities Policy **21**: 59–65.

Grimsey, D. and M. Lewis (2007). Public private partnerships: The worldwide revolution in infrastructure provision and project finance, Edward Elgar Publishing.

Grimsey, D. and M. K. Lewis (2002). "Evaluating the risks of public private partnerships for infrastructure projects." International Journal of Project Management **20**(2): 107–118.

Grimsey, D. and M. K. Lewis (2004). "The governance of contractual relationships in public–private partnerships." Journal of Corporate Citizenship **15**: 91–109.

Huang, C.-C. and T.-L. Tseng (2004). "Rough set approach to case-based reasoning application." Expert Systems with Applications **26**(3): 369–385.

Hwang, B.-G., X. Zhao and M. J. S. Gay (2013). "Public private partnership projects in Singapore: Factors, critical risks and preferred risk allocation from the perspective of contractors." International Journal of Project Management **31**(3): 424–433.

India, G. O. (2023). Retrieved 2 December 2023, from https://www.indiabudget.gov.in/.

Iyer, K. C. and Jha, K. N. (2005). "Factors affecting cost performance: evidence from Indian construction projects." International Journal of Project Management **23**: 283–295.

Jiang, A., Y. Zhang and Y. Ao (2022). "Constructing inclusive infrastructure evaluation framework—analysis influence factors on rural infrastructure projects of China." Buildings **12**(6): 782. https://doi.org/10.3390/buildings12060782

Kim, G.-H., S.-H. An and K.-I. Kang (2004). "Comparison of construction cost estimating models based on regression analysis, neural networks, and case-based reasoning." Building and Environment **39**(10): 1235–1242.

Kim, J. (2007). "A model and case for supporting participatory public decision making in e-democracy." Group Decision and Negotiation **17**(3): 179–193.

Kirkpatrick, C., D. Parker and Y.-F. Zhang (2004). Foreign direct investment in infrastructure in developing countries: does regulation make a difference? Transnational Corporations **15**(1): 143–171.

Kirschke, D. and K. Jechlitschka (2002). Angewandte Mikroökonomie und Wirtschaftspolitik mit Excel: Lehrbuch und Anleitung für eine computergestützte ökonomische Analyse, Vahlen.

Kirschke, D. and K. Jechlitschka (2003). "Interaktive Programmierungsansätze für die Gestaltung von Agrar-und Umweltprogrammen." German Journal of Agricultural Economics **52**(670-2016-45712): 211–217.

Kiryluk-Dryjska, E. (2013). "Fair division approach for the European Union's structural policy budget allocation: an application study." Group Decision and Negotiation **23**(3): 597–615.

Kiryluk-Dryjska, E. (2014). Formalizacja decyzji wyboru publicznego: zastosowanie do alokacji środków strukturalnych Wspólnej Polityki Rolnej UE w Polsce, Wydawnictwo Naukowe PWN.

Kiryluk-Dryjska, E. and P. Beba (2018). "Region-specific budgeting of rural development funds—An application study." Land Use Policy **77**: 126–134.

Kumari, A. and A. Kumar Sharma (2017). "Infrastructure financing and development: A bibliometric review." International Journal of Critical Infrastructure Protection **16**: 49–65.

Leruth, L. E. (2009). "Public-private cooperation in infrastructure development: a principal-agent story of contingent liabilities, fiscal risks, and other (un)pleasant surprises." Networks and Spatial Economics **12**(2): 223–237.

Levine, N. and Bland, N. L. (2010). Decentralization and Democratic Local Governance Programming Handbook. Center for Democracy and Governance Bureau for Global Programs, Field Support, and Research U.S. Agency for International Development Washington, DC 20523-3100 (available online).

Li, B., A. Akintoye, P. J. Edwards and C. Hardcastle (2005). "Critical success factors for PPP/PFI projects in the UK construction industry." Construction Management and Economics **23**(5): 459–471.

Mapuva, J. and G. P. Miti (2019). "Exploring the uncharted territory of devolution in Zimbabwe." Journal of African Studies and Development **11**(12): 12–20.

Matsatsinis, N. F. and A. P. Samaras (2001). "MCDA and preference disaggregation in group decision support systems." European Journal of Operational Research **130**(2): 414–429.

Merna, T. and C. Njiru (2002). Financing infrastructure projects, Thomas Telford.

Mikesell, J. (2013). Fiscal administration, Cengage Learning.

Munyede, P. and J. Mapuva (2020). "Exploring public procurement reforms in rural local authorities in Zimbabwe." Journal of Public Administration and Governance **10**(1): 1–16.

Nishijima, K. and M. H. Faber (2009). "A budget management approach for societal infrastructure projects." Structure and Infrastructure Engineering **5**(1): 41–47.

NLC (2016). Paying for local infrastructure in a new era of federalism, National League of Cities.

OECD (2009). OECD Principles for Integrity in Public Procurement, ORGANISATION FOR ECONOMIC CO-OPERATION AND DEVELOPMENT.

OECD (2019). Capital budgeting and infrastructure, Paris, OECD Publishing.

Olusola Babatunde, S., A. Opawole and O. Emmanuel Akinsiku (2012). "Critical success factors in public-private partnership (PPP) on infrastructure delivery in Nigeria." Journal of Facilities Management **10**(3): 212–225.

Osei-Kyei, R. and A. P. C. Chan (2015). "Review of studies on the Critical Success Factors for Public–Private Partnership (PPP) projects from 1990 to 2013." International Journal of Project Management **33**(6): 1335–1346.

Outlook, G. I. (2017). Infrastructure investment needs 50 countries, 7 sectors to 2040. A G20 Initiative. Sydney, Global Infrastructure Hub.

Öztaş, A. and Ö. Ökmen (2004). "Risk analysis in fixed-price design–build construction projects." Building and Environment **39**(2): 229–237.

Parushina, N. V., N. A. Lytneva, M. E. Khanenko, M. G. Starostin and M. I. Klimoshenko (2020). "Budget risks in the management of small businesses in rural areas." IOP Conference Series: Earth and Environmental Science **548**(2): 022088.

Podger, A., T.-T. Su, J. Wanna, H. S. Chan and M. Niu (2018). Value for money: Budget and financial management reform in the People's Republic of China, Taiwan and Australia.

Pongsiri, N. (2003). "Public-private partnerships in Thailand: a case study of the electric utility industry." Public Policy and Administration **18**(3): 69–90.

Posner, P. L., S. K. Ryu and A. Tkachenko (2009). "Public-private partnerships: The relevance of budgeting." OECD Journal on Budgeting **9**(1): 1–26.

Rackwitz, R. (2000). "Optimization—the basis of code-making and reliability verification." Structural Safety **22**(1): 27–60.

Rath, R. and S. Samal (2015). "An impact of Foreign Direct Investment (FDI) On infrastructure development for the economic growth in India: an economic survey on Indian scenario." International Journal of Interdisciplinary Research **2**(5): 78–95.

Rozenberg, J. and M. Fay (2019) "Beyond the gap: How countries can afford the infrastructure they need while protecting the planet." Sustainable Infrastructure Series **1**, 6.

Sader, F. (2000). Attracting foreign direct investment into infrastructure: Why is it so difficult?, World Bank Publications.

Saha, D., S. H. Hong, T. Nair, A. Bhattacharya, E. Teo and S. Song (2021). Private participation in infrastructure (PPI): 2021 Annual Report, The World Bank.

Schmid, J. C., A. Hager, K. Jechlitschka and D. Kirschke (2010). Programming rural development funds–an interactive linear programming approach applied to the EAFRD program in Saxony-Anhalt. Structural Change in Agriculture/ Strukturwandel im Agrarsektor (SiAg) Working Papers 59523, Humboldt University Berlin, Department of Agricultural Economics, doi: 10.22004/ ag.econ.59523

Sharma, A. K. and E. Vohra (2008). "Foreign direct investment in the electricity sector: the Indian perspective." The Electricity Journal **21**(7): 63–79.

Sindane, J. (2000). "Public-private partnerships: case study of solid waste management in Khayelitsha-Cape Town, South-Africa." In L. Montanheiro and M. Linehan (eds.), Public and Private Sector Partnerships: The Enabling Mix (pp. 539–564). Sheffield: Sheffield Hallam University.

Sivak, M. (2017). "Has motorization in the US peaked? Part 9: Vehicle ownership and distance driven, 1984 to 2015." Report No. SWT-2017-4. Ann Arbor, MU: The University of Michigan, Sustainable Worldwide Transportation.

Stewart, T. J., A. Joubert and R. Janssen (2009). "MCDA framework for fishing rights allocation in South Africa." Group Decision and Negotiation **19**(3): 247–265.

Sutiyo and K. L. Maharjan (2017). District budgeting for rural development. Decentralization and Rural Development in Indonesia. Sutiyo and K. L. Maharjan. Singapore, Springer Singapore: 55–76.

Tas, E. and H. Yaman (2005). "A building cost estimation model based on cost significant work packages." Engineering, Construction and Architectural Management **12**(3): 251–263.

Terill, M. (2016). Cost overruns in transport infrastructure, Grattan Institute.

Thacker, S., D. Adshead, M. Fay, S. Hallegatte, M. Harvey, H. Meller, N. O'Regan, J. Rozenberg, G. Watkins and J. W. Hall (2019). "Infrastructure for sustainable development." Nature Sustainability **2**(4): 324–331.

Tinbergen, J. (1952). On the theory of economic policy. Books (Jan Tinbergen). Retrieved from http://hdl.handle.net/1765/15884

Touran, A. (2003). "Calculation of contingency in construction projects." IEEE Transactions on Engineering Management **50**(2): 135–140.

Utt, R. D. (1999). "How public-private partnerships can facilitate public school construction." In the Domestic Policy Studies Department (Ed.). (pp. 4–6). Washington DC., The Heritage Foundation.

Wang, S. Q., R. L. Tiong, S. K. Ting and D. Ashley (1999). "Political risks: analysis of key contract clauses in China's BOT project." Journal of Construction Engineering and Management **125**(3): 190–197.

Widdus, R. (2017). Public–private partnerships for health: their main targets, their diversity, and their future directions. Global Health, Routledge: 431–438.

Wojewnik-Filipkowska, A. and D. Trojanowski (2013). "Principles of public-private partnership financing – Polish experience." Journal of Property Investment & Finance **31**(4): 329–344.

Xenidis, Y. and E. Stavrakas (2013). "Risk-based budgeting of infrastructure projects." Procedia – Social and Behavioral Sciences **74**: 478–487.

Yu, W. D. (2006). "PIREM: a new model for conceptual cost estimation." Construction Management and Economics **24**(3): 259–270.

8 Infrastructure governance in rural and regional contexts

8.1 Introduction

While rural development is closely linked to the UN's sustainable development goals (SDGs) across many countries, discussions or evidence on the context-specific infrastructure planning and implementation for achieving the SDGs targets are not quite widespread in the mainstream literature. Often the topic of infrastructure is confined to the urban centres due to the money, power, and overall business prospects involved in infrastructure investments. Usually, rural infrastructure investment is seen as the government's responsibility and implementation is often done by the public or government organisations without the required skills, training, competencies, and governance framework. The concepts such as project-based, time-bound, and cost-effective implementation of infrastructure projects for providing need-based services for rural communities are compromised due to lack of appropriate governance models. Following the discussions in the preceding chapters on the vast dimensions of infrastructure matters in the context of over 40% rural population in the globe, one of the key necessities for infrastructure implementation is good governance. Infrastructure is a broad term and the scale varies from type, size, sector, and socio-economic standards of the project locations. Infrastructure and governance in the context of the current book is about service delivery and empowerment of rural communities through Smart Villages. Infrastructure systems are interrelated and so as the rural needs and requirements. While implementation of SDGs in a context-specific manner has the potential to impact rural lives and uplift rural conditions by providing accessibility to fundamental needs, a piece-meal or fragmental approach to infrastructure is unlikely to deliver the outcomes due to the following reasons:

Firstly, individual effort or a single piece of infrastructure in a particular sector does not create any impact on the community's well-being. For instance, investment in a single rural road does not provide necessary support for agriculture or power needs. Similarly, building a physical school building does not necessarily impact students' access to quality education unless resources are available for teachers' recruitment and training, sports and library facilities for students, uninterrupted and affordable power supply,

DOI: 10.1201/9781032622323-8

basic health access, healthy diet, comfortable housing and shelter including resilience again risks eventualities and uncertain conditions, etc.

Second, infrastructure investment and implementation from one community to the other may be different and that depends on the need-based priorities in localised settings. Thus, understanding of the demography, real-time assessments of the needs, requirements, and potentials among all sections of the local community, and adaptive and participatory bottom-up governance model are the key ingredients for driving a successful infrastructure programme and making a tangible impact required to create the smart villages.

Drawing the challenges, discussions and deep insights on numerous interrelated aspects involved in rural development from the previous chapters, this last chapter of this book aims to put forward a comprehensive governance model for creating smart villages based on the author's long research and scholarly contributions in the related fields.

8.2 Significance of good governance in rural infrastructure

Good and well-managed governance is important for effective infrastructure implementation and service delivery. However, good governance is a big question as no universally accepted governance works in every situation. Governance is kind of a culture as its forms and characteristics vary from one nation to another (Beer 2014). Figure 8.1 depicts the key elements of a typical governance model for developing infrastructure in rural settings. As illustrated, there are eight key elements namely, (1) *need-based planning and target setting*, (2) *transparency and accountability,* (3) *governed by the laws*, (4) *regulatory framework*, (5) *participatory design, integrated value*, (6) *heritage and culture*, (7) *proactive government*, and (8) *review, monitoring, and control*.

Given the scarce rural funding, need-based planning and target-setting ability is one of the key requirements for effective governance. Need-based planning for infrastructure results in a shorter lead time for projects to plan, execute, and operate for bringing direct benefits to the communities in a shorter timeframe. The operational performance of the project can easily be

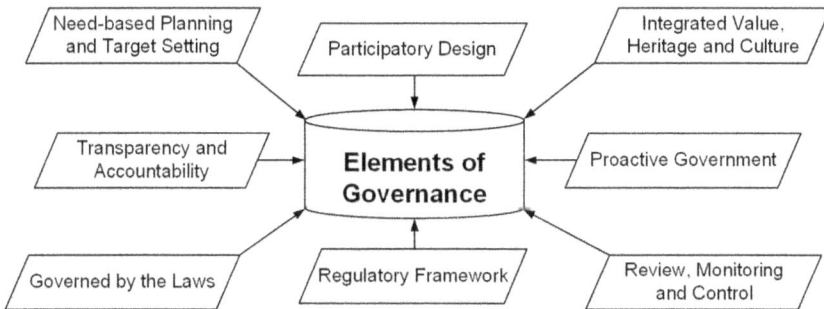

Figure 8.1 Governance model for rural infrastructure implementation.

linked to the target outputs and assessed against the expected thresholds. Transparency and accountability are crucial elements in governance not only to make the policymakers and executing organisations accountable for their responsibilities but also to generate trust and confidence among the citizenry in the region (Blair 2000). The use of public funds and prioritisation of projects based on citizen's needs make the overall outcomes acceptable to the community which eventually results in increased usage and utilisation of project facilities among the public. Rule of law and underlying rule-based governance are also important for both proponents and beneficiaries on the same page of infrastructure development and operations. Rule of law in governance models allows wider accessibility and participation for relevant parties and developing meaningful partnerships in the infrastructure projects.

Good governance must integrate the community's values, heritage, and culture in the design and development of infrastructure projects. Authorities need to be mindful and sensitive about the community's values, heritage, and culture and must ensure that the infrastructure projects developed in a particular location do not destroy but retain and reflect for making the facilities inclusive in a societal context. For instance, local fauna and flora, local craftsmanship, and local materials including retention and maintenance of natural ecology are some of the key considerations when designing projects in a particular location. Participatory design is one of the most powerful concepts for giving a voice to the citizenry and enticing them to participate in projects. Participatory design or co-design involves citizens' direct input of the core values of the community including direct interventions in addressing the gaps and necessities arising from lack of infrastructure provisions (Blair 2000). With the reasonably cheaper accessibility to technology and increasing bandwidths of wireless broadbands, comprehensive demographic data and necessary smart data analytics using a myriad of available tools make the participatory process easy and convenient in many settings. Examples of smart data collection processes and underlying data analytics used in the Authors' research are discussed in a later section of this chapter.

Regulatory frameworks such as stipulated quality standards, risks and uncertainty management, adaptation of workers' unions, or compliances and certifications by relevant standards are all very important in good govern- ance. This ensures not only high standards in project facilities but also transferability and versatility of the project functionalities for easy integra- tion of other infrastructures as necessary. A proactive government with underlying democratic values and objectivity in providing services to the citizens is crucial for maintaining good governance in all settings. As discussed in Chapter 7, the decentration of power and local representations at the community end in the democratically elected government system is considered effective for promoting proactive governance. Finally, close monitoring and review processes of the infrastructure interventions by independent authorities including assessments of the effectivity of the infrastructure functionalities in post-construction and value for money in both financial and non-financial terms are important for continuous improvement of services and empowering the community. Some of these elements are further discussed with examples throughout the chapter.

8.3   Reviews of governance practices – A few selective examples

In democratic government systems, one of the key methods for allowing citizens' voice in the governance system is by ensuring representative members and participation in the general elections from the community or community groups. Examples of the community groups may include women groups, indigenous and tribal groups, or other minority sections within the society. Examining the examples from Bolivia, the Philippines, Mali, and India, Blair (2000) asserted that the schemes involving community representation through elected local members as community voices were only partially successful in many instances. Most of the time, local elites control the election process by putting forward their preferred representatives which undermines the true voice of the minority groups within the community. In some instances, wives of the husbands of the elite groups would get elected and husbands will influence the power as per their interests. There are examples where, often prominent personalities and vocal members among the indigenous groups would be offered employment in sectors controlled by elite members so that they can control or have a say through such representations.

As highlighted by Crook (2003), in the case of Zimbabwe where poorer communities had been provided with protective or reserved seats in the local government elections, there were protests by the opposition parties (Crook 2003). The ground for the protest was that poorer communities lacked the education and intellectual power to become representative members and thus inappropriate to participate in the election process. Promoting a universalistic principle for representative governance, Blair (2000) suggested an idealistic approach for benefitting all sections of the community through education. He suggested that existing schools across the board or schools with needs in certain communities may be provided with public funding with a target of a high literacy rate among the children for the betterment of the community in the long run. However, the possibility of siphoning the funds by the elite groups is feared to be still a reality in most developing countries. One of the other alternatives for a potential win-win situation among both elite and poorer sections of the community is the campaign for "primary education for all" as a long-term goal with the possibility of getting unconditional support without bias (Blair 2000).

Figure 8.2 illustrates a typical flow of public funds in India within a decentralised governance system. Being a vast country with a population of over 1.4 billion residing across 28 states and eight union territories, India perhaps can't function with a fully centralised or even two-tier Federal-State system. From the population point of view, many of the Indian states are larger than many countries in Europe and also Australia continent. Thus, the

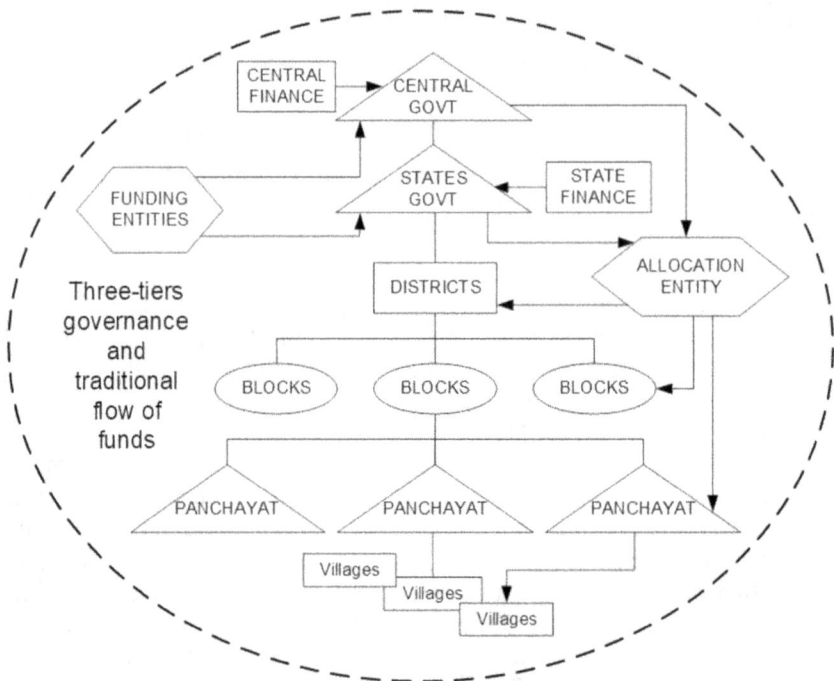

Figure 8.2 Three-tiers governance and flow of public funds in India.

power-sharing structure from central government to state and then state to local government is one of the core elements within the decentralised model. A decentralised model is proven to be very effective for managing and settling local issues at the grassroots level. People within the local community are usually knowledgeable and with easy access to the local authorities, many of the local issues could be resolved within a short timeframe. At the top of the decentralised pyramid illustrated in Figure 8.2, the central government holds the maximum power and financial capacity. The second level is the state governments where in addition to the share from central, independent financial capacity exists from local and lateral funding sources as discussed in the previous chapter. The third or bottom level in the pyramid is the local government which is also known as Panchayat Raj or Gram Panchayat. Gram Panchayats are the rural local government comprising wards or villages. Many Panchayats may combine to make a Block. A district may comprise a few blocks representing the functional responsibilities of the decentralised model. That means the local government structure comprises villages, panchayats, and blocks up to a district. Gram panchayat also known as council is an elected body with representative members drawn from wards or villages. The council is usually the sole decision-making body for a particular village within its jurisdiction and the funds usually come from state and central allocations stipulated by the specific schemes. Elected members of both central and state constituencies are also members of the local council of the specific district. As the councils are responsible for village administration, semi-urban areas and big cities are governed by municipal corporations headed by an officer equivalent to the mayor in many Western countries.

As the administrative power of gram panchayat lies with the elected members of the council and respective office bearers, there are inefficiencies in the utilisation of funds to the best value and interests from the community perspective. Often the governing members in the councils do not keep up with the required skills and training and lack the capacity for holistic implementation of the project. Vested interests in the election process among a particular group of people as mentioned above may still be the case where public money is utilised effectively for improving required services in the community. Objectivity in citizen participation and integration of community viewpoints in grassroots-level decision-making on infrastructure projects are often compromised with top-down dictatorial funding regimes which flow either directly from the central government or via state and district administrations. The bad implementation of the flagship programme from the Central Government "The Pradhan Mantri Awas Yojana- Gramin (PMAY-G)" with a grand target of "Housing for All" is a classic example where a top-down funding scheme couldn't deliver the best interests of the community in the current governance practice (India 2014, Dhanabhakyam and Shobanageetha 2018). One of the key issues with the PMAY-G scheme is the misalignment of the community's needs at the individual or beneficiary level with the housing design and construction and most often the unfinished end-products due to lack of funding or cost escalations (Dhanabhakyam and Shobanageetha 2018, Doloi 2022).

8.4   Data-driven decision-making for participatory governance

A strong relationship exists between the community, local culture, utilisation of public funds, and effective governance. Examining the way rural funds are being used in the case of Indonesian decentralised fiscal responsibility, Pal and Wahhaj (2017) asserted a significant relationship between the local culture and funding effectiveness for delivering goods and services through relevant infrastructure projects. The local leader in the village community had a much stronger influence in the informal governance model to highlight the need for local road connectivity with neighbouring villages for improving local trades and services which were successfully implemented. This concept of informal community leaders with the ability to influence public policies is also quite common in other countries such as India. In remote villages, usually community is uninformed or not quite vocal about bringing their concerns or issues onto the surface and making claims publicly. But with the help of informal leaders, they can express their concerns even colloquially and then get their voices heard to make an impact in the decision-making process of the government. As suggested by Beer (2014), the government-appointed community task force is another common alternative for developing partnerships between the government and the community and collecting community voices for informing policymaking. However, there is a concern that government representatives in the task force may be perceived to be biased and may push the government agenda by manipulating the community's collective concerns (Beer 2014). If in reality task force can work in collaboration with the community, the effectiveness of the partnership could grow with a potential for getting public policies fully aligned with the community's expectations.

The example of the LEADER programme in many countries in Europe relies on the local action groups to represent the vast majority of the community in the rural development regime (European Network for Rural Development 2016). The LEADER programme is considered successful due to the ability to include voices of the people with genuine interests who wish to make an objective difference in local development strategies. While the success of the LEADER programme still relies on the willingness to partner with both political parties and the community, political parties with a majority in the parliaments tend to support more than the minority or coalition partners. The political majority usually drives for greater accountability in public policy making and hence community partnerships in such situations are considered more conducive for devising collaborative strategies in rural development projects (Falkowski 2013).

As seen, community partnership and participatory governance model is a proven concept for effective implementation of rural development projects. However, key challenges lie in the collection of empirical evidence that reflects the true concerns of the community. The decisions taken on projects based on only correct facts can make a real impact and support the community with the required services. Thus, the success of the participatory governance model relies on the ability to identify objective representations so that the voices of the community are reflected at large. The following section briefly discusses how technology-enabled smart data platforms and data analytics were applied in a research project for collecting data from the target community in entirety and data analytics supported reflecting the community concerns in totality.

8.5 Smart data framework and data analytics

Smart data refers to the technology-supported real-time comprehensive demographic data collected on the community at the household level. The comprehensiveness of such a data set is a key that can provide a snapshot or X-ray of the community and highlight the critical indicators that are necessary for decision-making on context-specific interventions.

In a research project conducted by the author at the Smart Villages Lab (SVL), a purpose-built computer application (app) was developed to collect

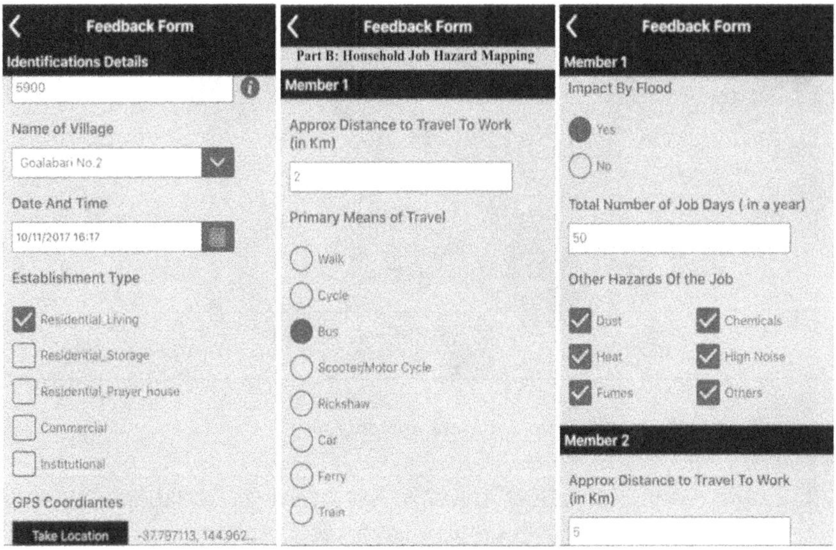

Figure 8.3 Smart data collection application.

demographic data from rural communities using computer Tablets or Phones. Figure 8.3 shows an interface of the app installed on a smartphone. After receiving due ethics approval from the University of Melbourne, the app was deployed across 37 rural villages in the river island, Majuli located in the North-Eastern state of Assam in India. Six volunteers were recruited from the local community to provide necessary training and support to assist in collecting socio-economic data from over 2,500 households across the region. The questionnaire was over 20 pages long, comprising some extensive questions to capture all facets of information on every member within the households. With a highspeed 4 G network, the data was then transmitted onto the cloud storage which was then accessed at the SVL for conducting necessary computations and generating analytics. As part of the data collection, every single household was photographed which was later geo-referenced and attached to the household data in a Smart Data Platform.

Figure 8.4 shows an interface of the Smart Data Platform developed as a proof-of-concept in the research project. The Smart Data Platform facilitates rapid assessment of the community's socioeconomic conditions and thereby identifies the needs by analysing data from the field survey data collected by using the purpose-built app using smartphones.

The prototype software developed as Smart Data Platform demonstrates multiple governance support features such as:

• Finding the most relevant data when it's wanted by providing data selection and conditional filtering at state, district, village, and house levels.

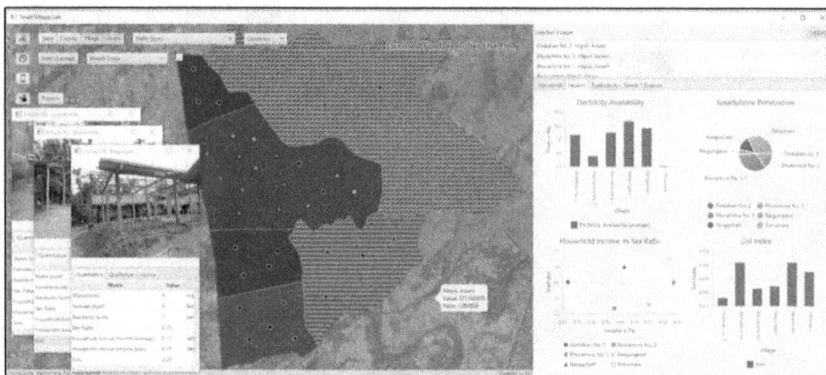

Figure 8.4 Smart Data Platform and data analytics.

- Allocating resources and effort effectively through collaboration with government departments with integrated project overlays.
- Answering the right questions by offering highly configurable editing and addition of analysis metrics.
- As the community and demography change in the development processes, adding new data in the Smart Data Platform is necessary for:
  - Rapid and comprehensive assessment of the community integrating new information including any unstudied areas, leveraging the Survey App and powered by the automated machine learning techniques.
  - Seamless incorporation of field data for continuous data analytics for examining the new areas of interventions and decision-making for rapid implementation.

The real-time and georeferenced feedback functionality within the Smart Data Platform allows the communities to be proactively engaged by:

- reporting of problems by community members, such as worsening flood conditions, dilapidated roads, broken down farm irrigation systems, etc.;
- photographic reporting of the on-the-ground conditions through detailed household data including individual house photos; and
- developing self-reliance, trust, and confidence including accountability and transparency in solving local problems.

Overall, the Smart Data Platform concept offers a flexible and powerful advantage to policymakers over infrequently-collected data sources, or unstructured, informal knowledge-gathering processes. This adds a significant capability in rural governance or co-governance for effective implementation of the development programme integrating the context-specific conditions, ensuring tangible outcomes, and meeting the targets.

8.6 New generation governance framework for rural infrastructure implementation

As discussed throughout the book and particularly in Chapter 5, the key argument put forward in this book on rural infrastructure is about the context-specific implementation of the UN's SDGs under the auspice of the Smart Villages programme. The author argues that while SDGs are the lowest denominations for many of the rural communities in receiving the basic minimum necessities, a blanket policy for implementing SDGs in any rural setting does not necessarily result in upward trajectories for empowering the communities in a sustained manner. Rather, a data-driven and evidence-based approach is required enabling priority settings on goals and targets including underlying infrastructure provisions for purposeful implementations.

Referring to Table 5.2 in Chapter 5, to materialise the collective benefits of SDGs, assessment of the required functionalities from the interconnected infrastructure systems is the important first step in the decision-making process within the governance structure. The governance structure not only requires the capability for accurate identification of the needs and extent of the infrastructure provisions but also exercises the power of influence among the key stakeholders for the successful execution of the plans aligning the objectives and meeting the targets. Figure 8.5 illustrates a comprehensive governance framework for rural infrastructure implementation. As seen, at

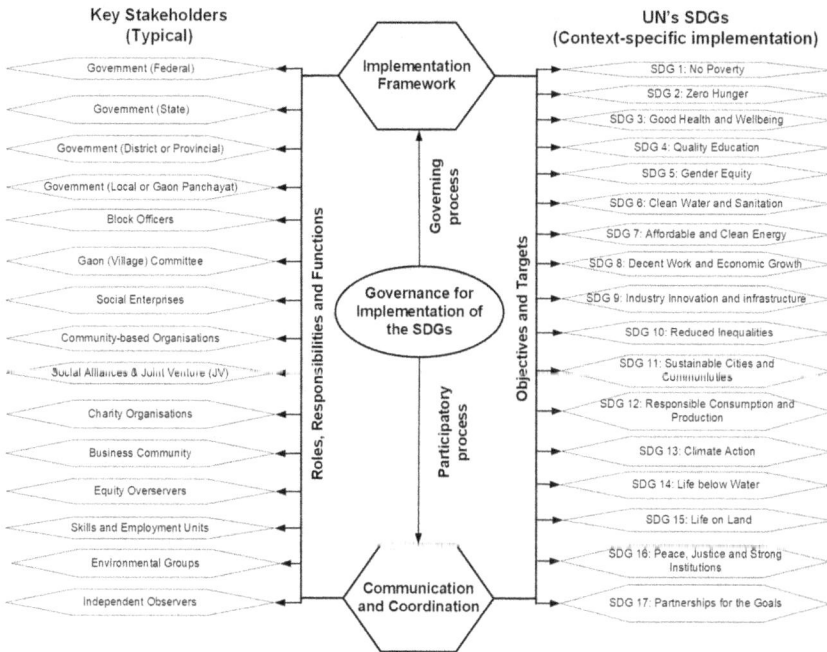

Figure 8.5 A new generation governance model for rural infrastructure implementation.

the core of this new generation governance framework are the two key elements, *key stakeholders* and *the UN's SDGs*. While the key stakeholders are responsible for the context-specific implementation of the SDGs in any setting, the two most crucial enabling forces are the *coordination and communication* and governing processes or strategies for *effective executions*. As much as the objectives and targets of the SDGs are important for setting a clear roadmap for development planning, so is the importance of the roles, responsibilities and functions associated with each stakeholder in both individual and relative terms. Communications and coordination required among the stakeholders for collective implementation of the SGDs are important to examine in the planning process so that execution can be carried out effectively keeping the plans intact (Images 8.1 and 8.2).

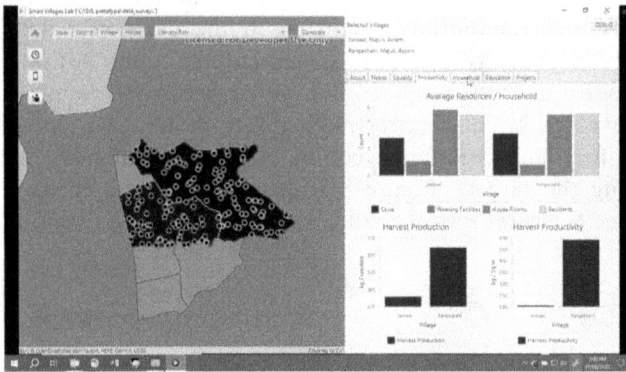

Image 8.1 Example of data analytics in action for village-level assessments of average resources per household and harvesting productivity.

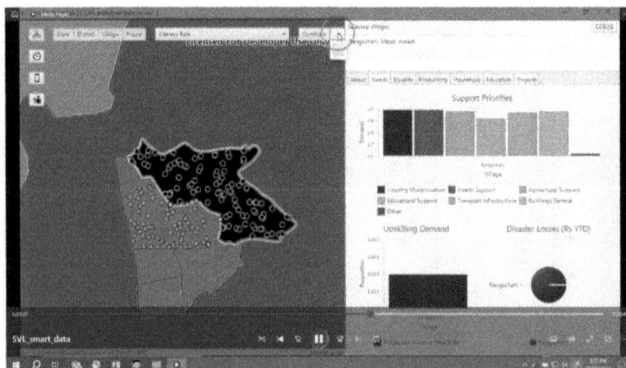

Image 8.2 Example of data analytics in action for village-level assessments of support priorities, upskilling demand, and disaster losses.

8.7 Responsibility sharing among the stakeholders in SDG implementations

Referring to Figure 8.5, coordination and communication among the key stakeholders across multiple levels is one of the key elements in the governance model. Stakeholders are the people or organisations we are directly or indirectly involved in the project and either can impact or be impacted in the course of the project. Stakeholders whether on the proponents' side or recipients' end usually have formally or informally defined sets of roles, responsibilities and functions in the project. The implementation of the SDGs for community development programmes involves a large number of stakeholders with varied roles, responsibilities and functions.

Table 8.1 depicts a list of 15 stakeholders along with their roles and functions involved in a typical project. Varied levels of coordination and communications among the stakeholders occur about the SDGs and their specific and collective implementation.

Figure 8.6 highlights the responsibility sharing between stakeholders concerning the scope and activities associated with the implementation of each SDG. This network-based association between the stakeholders and SDGs is based on the binary or weighted input usually provided by the experts involved in the implementation of similar projects (Doloi 2012, Doloi, Pryke et al. 2016). The mapping tool, Social Network Analysis (SNA), which is a widely used method, especially in the sociology-based research domain, is then used to study the network relations. Through his extensive research, the author has demonstrated the use of SNA for studying the coordination and communication in the network structure in various application contexts (Doloi 2012, Doloi, Pryke et al. 2016).

Over past decades, the use of SNA as a research method has become quite popular among researchers for studying the complex interactions involved in project development processes (Mandarano 2009, Lyles 2015). The SNA method facilitates the network analysis among the actors or instances about a range of network characteristics such as Degree Centrality, Eigenvector Centrality, Closeness Centrality, Betweenness Centrality, etc. (Doloi 2012). There are many examples of how the researchers have been using SNA in examining projects or project processes. Focusing on an environmental planning project in the United States, Lyles (2015) revealed the close interactions between the planners and the planning processes (Lyles 2015). His research suggested that a stronger association between the planning processes and planners' roles depicts better professional services in practice. Stronger and more frequent communications among the stakeholders can exercise much stronger power and influence on project issues. Network-based relationships and underlying network characteristics can be used for developing social capital and achieving significant collective capacity in the project implementation process (Mandarano 2009, Doloi 2018).

As seen in Figure 8.6, two different nodes, round and square nodes, represent stakeholders and SDGs respectively. The sizes of the nodes are by

Table 8.1 Typical stakeholders, their roles, and functions

Stakeholders (typical)	Roles	Functions
S1: Govt (Federal)	Government	• Policymaking (national) • Budgeting and funding
S2: Govt (State)	Government	• Policymaking (states or provinces) • Costing, budgeting, and funding
S3: Govt (District Government)	Government	• Funding administration • Policy execution and compliance • Coordination and communication
S4: Govt (Gaon Panchayat)	Local Government	• Project execution • Coordination and communication • Funds monitoring and control • Progress monitoring and reporting • Community liaison
S5: Block Officer	Local Government	• Project execution and monitoring • Reporting • Community liaison
S6: Gaon Committee	Beneficiary	• Project proposals and demands • Monitoring and reporting • End users and development
S7: Social Enterprises	Social entrepreneurs	• Job opportunities and skills • Local business opportunity • Local resources and engagement
S8: Community-based Organisations	Local Executers and task-force	• Coordination and communication • Local resources and engagement • Community well-being and development
S9: Social Alliances and Joint Ventures	Partnerships	• Job opportunities and skills • Business partnerships and supply chain • Local engagement
S10: Charity Organisations	Non-government organisations	• Community wellbeing • Equity and accessibility • Community support
S11: Business Community	Business entrepreneurs	• Job opportunity • Income generation • Business growth and development
S12: Equity Observers	Compliances and support	• Equity and accessibility observers • Community liaising • Compliances

(*Continued*)

Table 8.1 (Continued)

Stakeholders (typical)	Roles	Functions
S13: Skills and Employment Unit	Business and community support	• Skill development • Education and training • Minority groups' empowerment
S14: Environmental Group	Compliances and support	• Environmental Compliance • Ecology and natural conservations
S15: Independent Observer	Compliances and support	• Compliances • Assessments, monitoring, and reporting • Whistleblower

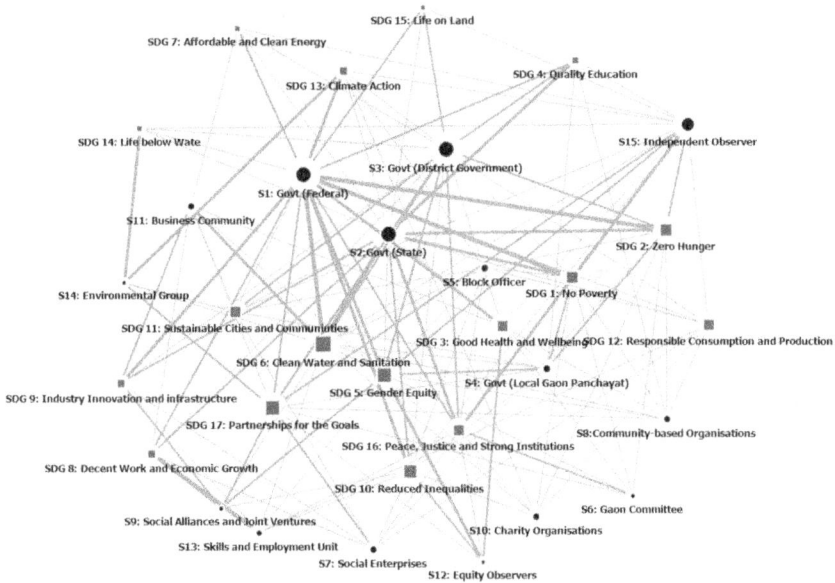

Figure 8.6 Responsibility sharing network of stakeholders in SDGs the implementation.

the Degree Centrality (DC) measure, which is one of the key network characteristics depicting the degree of connectedness with the other nodes in the network structure. The higher the size of the node, the higher the connection points with the other nodes. The arrows and thickness of the lines between the nodes represent the direction and amount of information flow respectively. This network representation between stakeholders and nodes eventually highlights the relative importance of the stakeholders and the

SDGs in the implementation process. Similar network analysis using other SNA measures, such as Eigenvector Centrality can be used for understanding the power of influence of one node over others in the overall network structure (Doloi 2018).

Visual depictions of information flow between the stakeholders and the SDGs provide significant insight for the policymakers in exercising context-specific governance models in the implementation process of SDGs in practice. Extending the discussion further, the following section will delve into the processes of examining the stakeholder-stakeholder communication and coordination on the implementation of each of the SDGs in an individual context.

8.8   SDG-specific coordination and communication among stakeholders in project implementation

Following the discussions on collective implementation of SDGs using the new generation governance model in the above sections, this final section of the project governance delves into examining the roles and responsibilities of the key stakeholders for context-specific implementation of each SDG. Every SDG is different in scope, costs and budget, infrastructure requirements and interventions for achieving the targets. Demographic and socio-economic data collected on the community-specific environment and then processed through the Smart Data Platform (see Section 8.5) provide the needs and priorities of the specific SDG and its implementation. The roles and responsibilities of the key stakeholders involved in the implementation of the specific SDG can be mapped using binary or weighted scores as discussed earlier. Using the SNA tool, the stakeholder-stakeholder association data matrices in computed and the underlying network characteristics are then used for examining the coordination and communication among stakeholders in each SDG context. Table 8.2 presents the SDG-specific coordination and communication network of stakeholders in the implementation process.

Referring to the fifteen stakeholders listed in Table 8.2, communication and coordination required among the stakeholders for the implementation of each of the seventeen SDGs were analysed based on the weighted association matrices. As mentioned above, the weighted or binary association matrices are usually developed based on the input from experts involved or experienced in similar projects. As seen, stakeholders' coordination and communication networks are placed against each SDG following the same order from Table 5.2 presented in Chapter 5.

One of the prevailing network characteristics, Eigenvector Centrality (EC) which is a measure of the "power of influence" of a particular node concerning the other nodes in the network, is used to size the stakeholders' nodes about each SGD (Doloi, Pryke et al. 2016, Doloi 2018). The directed arrows show the flow of information from one node to the others and the thickness of the arrows depicts the amount of information exchanged (the thicker the arrow, the more amount of information being shared and

Table 8.2 SDG-specific coordination and communication network of stakeholders in the implementation process

Stakeholders coordination and communication

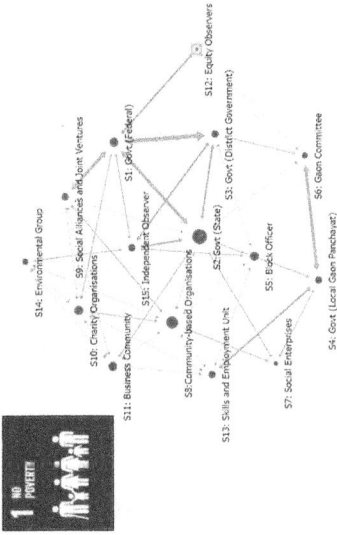

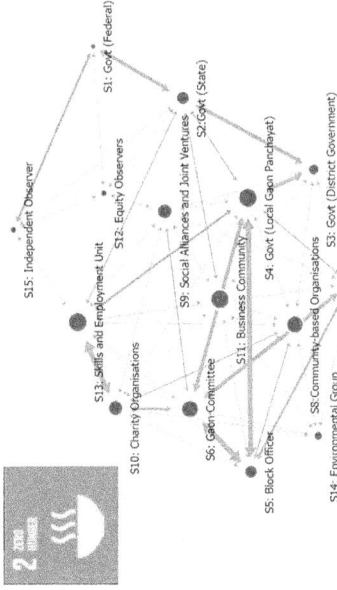

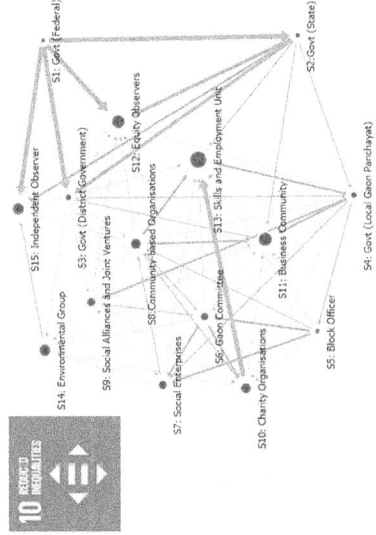

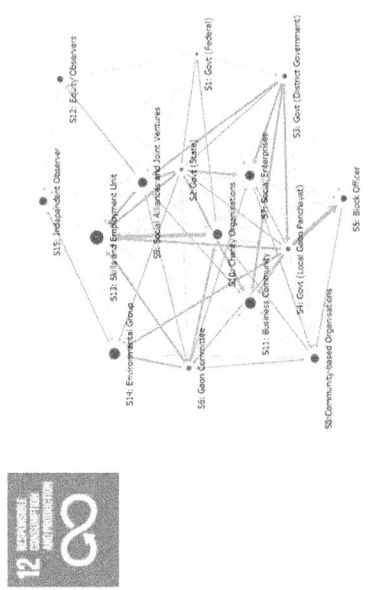

Stakeholders coordination and communication

(Continued)

Table 8.2 (Continued)

Stakeholders coordination and communication

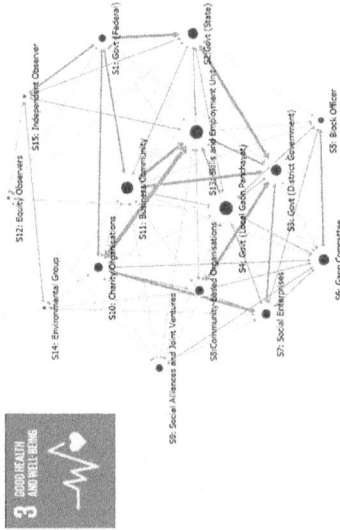

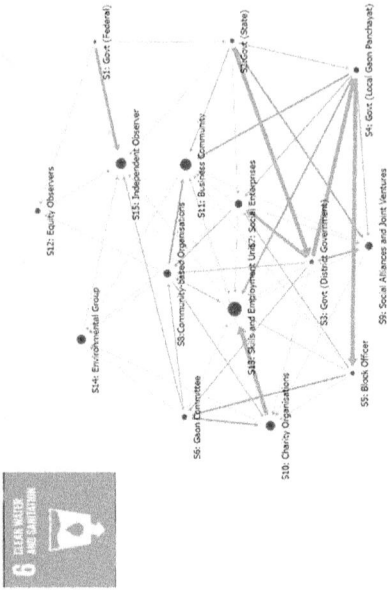

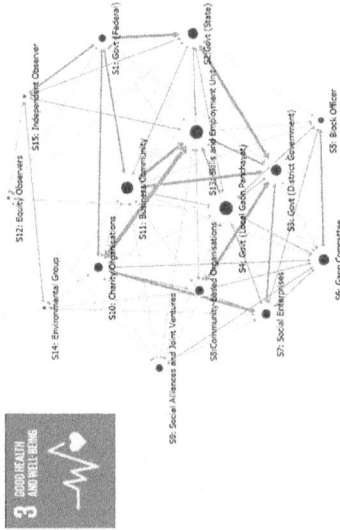

Stakeholders coordination and communication

(Continued)

Table 8.2 (Continued)

Stakeholders coordination and communication

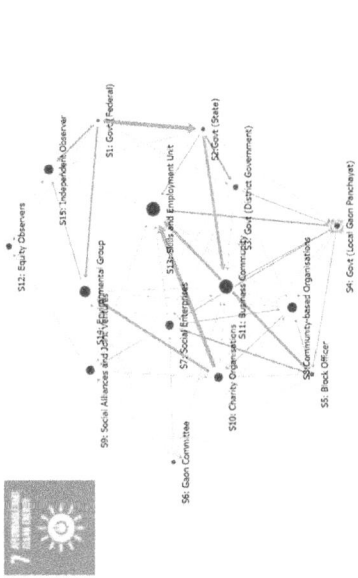

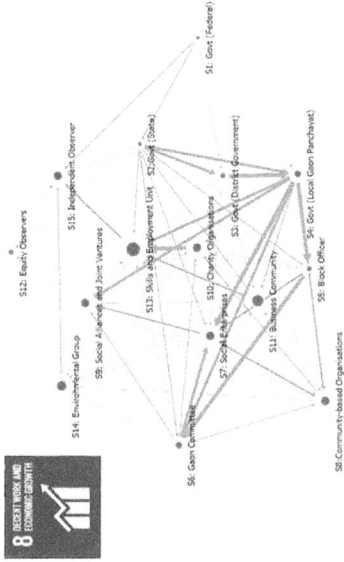

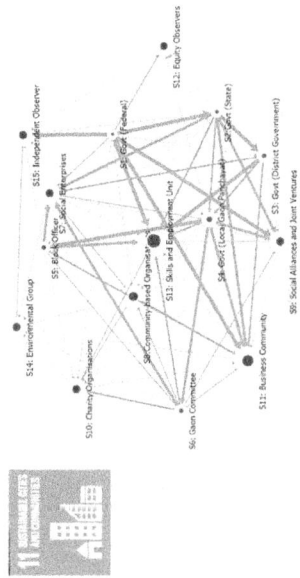

Stakeholders coordination and communication

(Continued)

Table 8.2 (Continued)

Stakeholders coordination and communication	Stakeholders coordination and communication

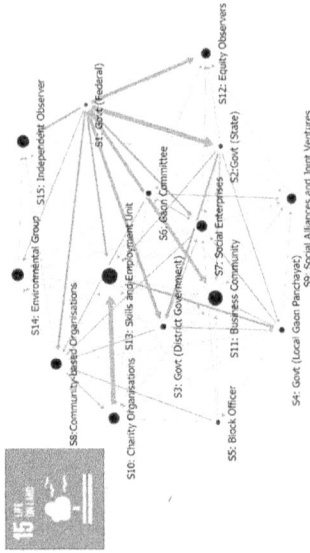

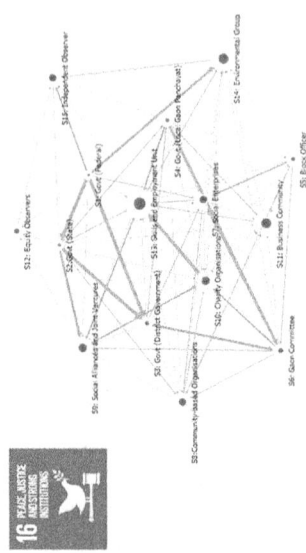

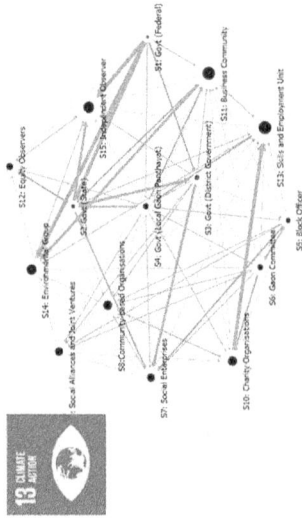

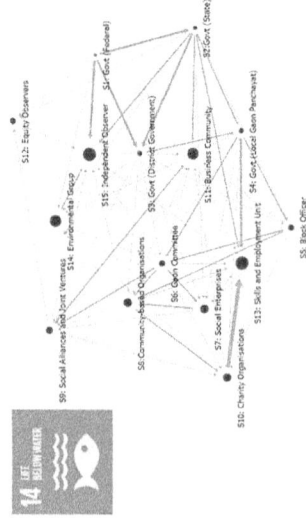

(*Continued*)

Table 8.2 (Continued)

Stakeholders coordination and communication	*Stakeholders coordination and communication*

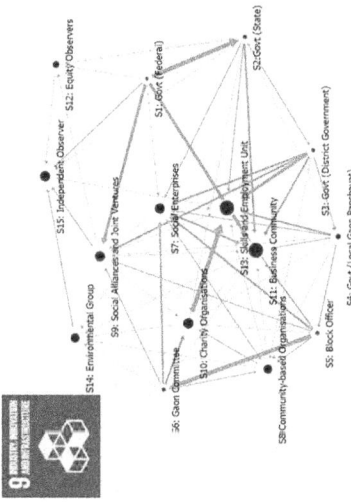

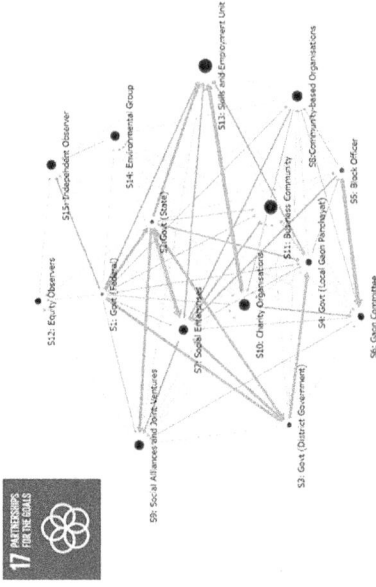

vice-versa) between the connected nodes. For instance, referring to SDG 1 (No Poverty), nodes S2: Govt (State) and S8: Community-based organisations are showing the two biggest nodes by the sizes which represent the nodes with the highest power of influence in implementing this SDG and meeting the target. By looking at the arrows' directions and thicknesses, node S1: Govt (Federal) is sharing the maximum amount of information with the three nodes S3: Govt (District Government), S2: Govt (State), and S9: Social Alliances and Joint Ventures. Similarly, the nodes with smaller sizes and the arrows with lesser thicknesses depict the stakeholders with lesser influence and a lesser amount of the information being shared in the process of implementing SDG 1 respectively.

A similar interpretation can be extended for all the SDGs by reviewing the individual network diagrams from the table. However, for the sake of brevity and to avoid repeated discussions, the analysis of the networks of the remainder of the SDGs is not included here.

Overall, comprehensiveness and accuracy in empirical data representing the target community in its entirety are highly crucial for the new generation governance model to act effectively. With the advent of rapid technological advancement and reasonably easy accessibility by the vast majority of rural communities, especially those residing in developing countries, the collection of empirical data has become less tedious compared to the practices a few decades ago. Thus evidence-based planning and implementation strategies, aided by the new governance framework presented in this chapter, should enable decision-makers to address the challenges being discussed in rural infrastructure development throughout the book. Data-driven, location-based planning, and context-specific implementation of infrastructure systems should assist in creating Smart Villages by cementing the "Rural Share" (referring to Chapter 5, Figure 5.3) of the 40% global population residing in rural conditions and contribute towards meeting the targets of the Sustainable Planet Initiatives (SPI).

8.9 Summary

In this final chapter, infrastructure governance was discussed from the rural infrastructure implementation. Having identified challenges and opportunities associated with the planning and development of rural infrastructure in the preceding chapters, this chapter brings together some of the unique thought processes from a governance perspective. There are numerous examples of governance models in practice but evidently, all models are not equally effective especially when it comes to rural development and community empowerment. Concepts such as citizen participation and bottom-up decision-making, community task forces are quite prevailing across many countries, yet effectiveness in delivering tangible outcomes and evidence of uplifting rural communities in entirety is rare. Voices of every section of the community a key to success, yet there is a clear lack of scientific approaches integrating everyone's viewpoints and then making inclusive public policies for purpose-built rural infrastructure provisions.

While the UN's SDGs provide a solid foundation for uplifting rural communities focusing needs and priorities of the population at the lowest denominations in the society, practices in holistic planning and delivery of the SDGs in a location-specific context are not widespread. Adopting a research-based approach, the author puts forward a few complementing concepts that entice participation as one of the core functions in the new generation governance model. The efficacies of data-driven decision-making for participatory governance powered by digital data framework and smart data analytics are discussed with research-based empirical examples. Finally, the network-based visualisations of stakeholders' interactions along with network-based characterisations of the relative significance of stakeholders' roles and responsibilities in the holistic implementation of the SDGs in both individual and collective contexts are presented. The data-driven and evidence-based governance model is demonstrated as a way forward towards creating Smart Villages through purpose-built infrastructure systems. Capitalising on the 40% population still residing in rural conditions, the necessity and ideology of the "Rural Share" is further augmented as one of the key contributing forces in the SPI and saving the planet.

References

Beer, A. (2014). "Leadership and the governance of rural communities." Journal of Rural Studies **34**: 254–262.

Blair, H. (2000). "Participation and accountability at the periphery: Democratic local governance in six countries." World Development **28**: 21–39.

Crook, R. C. (2003). "Decentralisation and poverty reduction in Africa: The politics of local central relations." Public Administration and Development **23**: 77–88.

Dhanabhakyam, M. and K. Shobanageetha (2018). "Rural development through pradhan mantri awaas yojana (pmay) in Coimbatore district." Journal of Emerging Technologies and Innovative Research **4**(5): 49–55.

Doloi, H. (2012). "Assessing stakeholders' influence on the social performance of infrastructure projects." Facilities **30**(11): 531–550.

Doloi, H. (2018). "Community-centric model for evaluating social value in projects." Journal of Construction Engineering and Management **144**(5). https://doi.org/10.1061/(ASCE)CO.1943-7862.0001473

Doloi, H. (2022). The usefulness of data analytics in Smart Villages development. The 5th International Conference on smart villages and rural development (Cosvard 2022). H. Doloi and A. Bora. Online, Smart Villages Lab (SVL): 82–90.

Doloi, H., S. Pryke and S. Badi (2016). The practice of stakeholders' engagement in infrastructure projects: A comparative study of two major projects in Australia and the UK. London, UK, RICS. **February:** 70 pages.

European Network for Rural Development 2016. LEADER/CLLD [Online] Available: https://enrd.ec.europa.eu/leader-clld_en [Accessed 19 Dec 2023].

Falkowski, J. (2013). "Political accountability and governance in rural areas: some evidence from the pilot programme LEADER+in Poland." Journal of Rural Studies **32**: 70–79.

India, G. O. (2014). "Pradhan mantri awas yojana- gramin." Retrieved 08/01/2024.

Lyles, W. (2015). "Using social network analysis to examine planner involvement in environmentally oriented planning processes led by non-planning professions." Journal of Environmental Planning and Management **58**(11): 1961–1987.

Mandarano, L. A. (2009). "Social network analysis of social capital in collaborative planning." Society & Natural Resources **22**(3): 245–260.

Pal, S. and Wahhaj, Z. (2017). "Fiscal decentralisation, local institutions and public good provision: evidence from Indonesia." Journal of Comparative Economics **45**(2): 383–409.

Epilogue

This book provided an overview of some of the little-understood and sometimes counter-intuitive best practices on rural infrastructure and value-based priorities in uplifting rural communities in developing economies. Drawing from the global literature and practice-based evidence across a complete spectrum of relevant disciplines, this book provided a clear articulation of the innovative ideas around harnessing the rural potential and empowering rural communities with added support in sustainable growth and progressive development in the context of interconnected infrastructure systems and improved living standards.

Governments in emerging economies, where rural poverty is most acute, have attempted to improve rural development through numerous public infrastructure schemes, especially in the water, power, and transport contexts. However, due to their lack of understanding of the intrinsic factors around community and individual priorities, such as potential, aspirations, value, heritage, and culture, these interventions have failed at multiple levels. Studies show that money allocated to public infrastructure for rural communities is often not spent effectively due to the attempted imposition of urban development techniques on rural areas. Assessing need-based infrastructure provisions in rural communities and understanding the design and development of interconnected infrastructure systems that meet community expectations is a difficult task. Meeting such challenges requires unique techniques and approaches that are sustainable, risk-averse, and resilient. These techniques and approaches have not yet been consolidated and mapped in the way that urban development approaches. This book acts as a guidebook to this new body of knowledge.

The author, coming from both a research and professional background in civil engineering, construction economics and infrastructure planning, procurement, and development, has a deep exposure and interest in rural life and has come to realise several things: First, the challenges associated with creating infrastructure provisions and need-based priority setting for supporting rural development need completely new approaches and innovations not necessarily same as the practices applied in cities; Second, the rural resilience, resourcefulness, lower carbon footprints, and community-spirit upon which such innovations are drawn, are distinctly different from the city-centric knowledge and requirements; Third, researchers and practitioners involved in rural work face challenges about how social and economic development "must" really be done, which not only meets the fitness-for-purpose but also contributes to the larger challenges of global warming and climate change.

There are endless possibilities for being distinctively innovative and finding community-led sustainable development practices within "The 40%" rural population, supporting the global effort to meet the "Net-Zero" target across the countries. Contrasting the current race for closing the "Urban-Rural Gap", contributions towards new and alternative development models for promoting "Urban-Rural Share" and contributing to the "Sustainable Planet Initiatives (SPI)" are mind-boggling and intellectually liberating across many interdisciplinary professions.

The United Nations (UN) Sustainable Development Goals (SDGs) provide a solid foundation for rural development and through a clear articulation in this book, the foundation is further expanded to support context-specific infrastructure provisions focusing on the rural communities and leading to the creation of Smart Villages.

Infrastructure needs for Smart Villages are unique and the principles for planning and development of sustainable infrastructure in rural settings are vastly different from the urban centres. While the term "sustainability" is widely used among researchers, practitioners, and policymakers including the general public, there is a lack of evidence of how varied sustainable practices can address context-based challenges in the infrastructure front and con-tribute towards broader sustainability targets. For the first time, this book has demonstrated how sustainable infrastructure development practices especially targeting over 40% of rural communities can lead to addressing global challenges such as climate change in a broader perspective. With the global population approaching 8.1 billion in 2024 and the unprecedented increase in demand for infrastructure development, community-led and context-specific development approaches powered by the new-generation governance model are a significant advancement of novelty being highlighted in this book. Throughout the book, a clear argument is put forward for rationalising the alternative sustainable development models by integrating vernacularisms of the rural community and building on thousands of years of local experience, and leading healthy, nature-based sustained lifestyles.

The author of this book argued that the extension of the urban-centric knowledge and practices (so-called brick-mortar development) in moder-nising 40% of rural communities is a gross violation of man-made acts against the sustainability outcomes of the planet. Instead, the knowledge, processes, and outcomes of alternative sustainable development models for rural infrastructure provisions among the rural communities will contribute in a significant way by containing the emissions of greenhouse gases and assist in meeting the net-zero targets not only nationally but perhaps at a global level which is being conceptualised as a Sustainable Planet Initiative (SPI) and a green mission within the Smart Villages Lab (SVL). This book comprising knowledge, examples, and concepts is certainly not an end product for addressing the rural infrastructure challenges and providing a complete solution. Instead, I hope this book will be viewed as the dawn of a new era for developing rural infrastructure for Smart Villages.

Index

Note: *Italicized* and **bold** page numbers refer to figures.

For Product Safety Concerns and Information please contact our EU
representative GPSR@taylorandfrancis.com
Taylor & Francis Verlag GmbH, Kaufingerstraße 24, 80331 München, Germany

www.ingramcontent.com/pod-product-compliance
Lightning Source LLC
Chambersburg PA
CBHW060255220326
41598CB00027B/4109